PRINCIPLES AND PRACTICE

Springer

Berlin
Heidelberg
New York
Barcelona
Hong Kong
London
Milan
Paris
Tokyo

Ian N. M. Day (Ed.)

Molecular Genetic Epidemiology – A Laboratory Perspective

With 28 Figures

Springer

Professor IAN N. M. DAY
Human Genetics Research Division
Duthie Building (MP 808)
Southampton General Hospital
Tremona Road
Southampton SO 16 6YD, UK

ISBN 3-540-41387-1 Springer-Verlag Berlin Heidelberg New York

Library of Congress Cataloging-in-Publication Data

Molecular genetic epidemiology : a laboratory perspective / Ian N.M. Day, ed.
    p.   cm. – (Principles and practice)
   Includes bibliographical references and index.
   ISBN 354041388X – ISBN 3540413871 (pbk.)
   1. Molecular epidemiology – Laboratory manuals.   2. Genetic epidemiology – Laboratory
manuals.   I. Day, Ian N.M.   II. Series.
   RA652 .M65 2001
   614.4 – dc21

Springer-Verlag Berlin Heidelberg New York
a member of Bertelsmann Springer Science+Business Media GmbH
http://www.springer.de
© Springer-Verlag Berlin Heidelberg 2002
Printed in Germany

The use of general descriptive names, registered names, trademarks, etc. in this publication does not imply, even in the absence of a specific statement, that such names are exempt from the relevant protective laws and regulations and therefore free for general use.

Cover design: D&P, Heidelberg
Typesetting: Best-set Typesetter Ltd., Hong Kong
SPIN 10713029   39/3130 – 5 4 3 2 1 0 – Printed on acid-free paper

Preface

This book covers selectively, a representative range of approaches in use in laboratories engaged in molecular genetic epidemiology. Following close on the tail of the Human Genome Sequencing Program is the study of the molecular genetic factors which make people and their diseases different. Intense interest is resulting in development of a bewildering array of theoretical and laboratory approaches. This volume presents overviews of core topics and techniques which are widely available and accessible to almost any researcher. It focuses on the robust, tried and tested methods, ranging through statistical planning, clinical sampling, microsatellite SNP and sequence calling and applications suitable for population studies in fields including immunogenetics, cardiovascular disease and subject identification.

Contributors are drawn from both an MRC Co-operative Group in Genetic Epidemiology (School of Medicine, University of Southampton, UK) and internationally in related areas of expertise and they are thanked both for responding to deadlines and patience in waiting for others to meet theirs.

Over the next decade, microarray chip and nanotechnology developments will flourish for ultra-high throughput applications, but will drive an increasing demand for established "medium" throughput and configurable approaches. We hope that this Volume will empower individual researchers in molecular genetic epidemiology.

June 2001   IAN DAY

Contents

4   Minisatellite and Microsatellite DNA Fingerprinting
PAUL G. DEBENHAM

5   Multiplex Polymerase Chain Reaction and Immobilized Probes:
Application to Cardiovascular Disease
SUZANNE CHENG

6   The Special Case of HLA Genes: Detection and Resolution
of Multiple Polymorphic Sites in a Single Gene
W. MARTIN HOWELL and KATHERINE L. POOLE

List of Contributors

XAIO-HE CHEN
Human Genetics Research Division, Duthie Building (MP 808), Southampton General Hospital, Tremona Road, Southampton SO16 6YD, UK
e-mail: ex@soton.ac.uk

SUZANNE CHENG
Department of Human Genetics, Roche Molecular Systems, Inc., 1145 Atlantic Avenue, Alameda, California 94501–1145, USA
e-mail: suzanne.cheng@roche.com

ANDY COLLINS
Human Genetics Research Division, Duthie Building University of Southampton, Southampton General Hospital, Tremona Road, Southampton SO16 6YD, UK
e-mail: arc@soton.ac.uk

IAN N.M. DAY
Human Genetics Research Division, Duthie Building (MP 808), Southampton General Hospital, Tremona Road, Southampton SO16 6YD, UK
e-mail: inmd@soton.ac.uk

PAUL G. DEBENHAM
Laboratory of the Government Chemist, LGC Building, Queens Road, Teddington, Middlesex TW11 0LY, UK
e-mail: pgd@lgc.co.uk

MICHAEL FITZGERALD
Genome Therapeutics Corporation, 100 Beaver Street, Waltham, Massachusetts 02454, USA

HEIDI GIESE
Genome Therapeutics Corporation, 100 Beaver Street, Waltham, Massachusetts 02454, USA

LESLEY J. HINKS
Human Genetics Research Division, Duthie Building (MP 808), Southampton General Hospital, Tremona Road, Southampton SO16 6YD, UK
e-mail: ljh3@soton.ac.uk

W. MARTIN HOWELL
Wessex Immunology Service, Southampton General Hospital, Tenovus
Laboratory, Tremona Road, Southampton University Hospitals NHS Trust,
Southampton SO16 6YD, UK
e-mail: m.howell@soton.ac.uk

CHENI KWOK
Exelixis Pharmaceuticals, Inc., 170 Harbor Way, P.O. Box 511,
South San Francisco, California 94083–0511, USA
e-mail: ckwok@exelixis.com

SANDRA D. O'DELL
Human Genetics Research Division, Duthie Building (MP 808), Southampton
General Hospital, Tremona Road, Southampton SO16 6YD, UK
e-mail: sdod@soton.ac.uk

KATHERINE L. POOLE
Combined Laboratory, Derriford Hospital, Plymouth PL6 8DH, UK
e-mail: katherinepoole@hotmail.com

KARIN SCHMITT
Exelixis Pharmaceuticals, Inc., 260 Littlefield Avenue, South San Francisco,
California 94080, USA
e-mail: kschmitt@exelixis.com

EMMANUEL SPANAKIS
Human Genetics Research Division, Duthie Building (MP 808), Southampton
General Hospital, Tremona Road, Southampton SO16 6YD, UK
e-mail: ms6@soton.ac.uk

Current address: Aventis Pharma, Evry Genetics Center, 2, rue Gaston
Crémienx, CP 5705, 91057 Evry Cedex, France
e-mail: emmanuel.spanakis@aventis.com

HANS-ULRICH THOMANN
Genome Therapeutics Corporation, 100 Beaver Street, Waltham,
Massachusetts 02454, USA

ANCA VOROPANOV
Human Genetics Research Division, Duthie Building (MP 808), Southampton
General Hospital, Tremona Road, Southampton SO16 6YD, UK
e-mail: mamvl@soton.ac.uk

KRISTEN WALL
Genome Therapeutics Corporation, 100 Beaver Street, Waltham,
Massachusetts 02454, USA

1 Mapping Genes for Common Diseases: Statistical Planning, Power, Efficiency and Informatics

ANDY COLLINS

1.1 Introduction

After a decade or so of progress with the identification of major genes, the emphasis has shifted towards genes for common diseases (complex traits). Real successes have been few so far. There has been a gradual appreciation of the difficulties presented by genes that have a relatively small individual phenotypic effect and which may show complex interactions. Furthermore, there may be different genetic determinants of the disease in different populations together with environmental effects and practical difficulties in obtaining sufficiently large samples of suitable material for analysis. At the same time, analytical methods have been in flux with many new tests and variations on existing methods appearing in the literature. The choice of analytical strategy is determined to a large extent by the nature of the disease and the DNA resources that can be obtained. Some assessment of power to detect a genetic locus can be useful; however, this may be of limited value given the ignorance of the nature of the genetic basis of the disease. Consideration of the efficiency of different approaches is also important and careful statistical planning should be undertaken.

Figure 1.1 gives an overview of the relationship between effect on the phenotype (measured by a single parameter, β) and disease allele frequency. There have been many successes with major genes (β of 1.5 or greater), rare alleles which have a large phenotypic effect. Examples of such single gene disorders are the cystic fibrosis gene (CFTR) (Kerem et al. 1989) and Huntington's disease (HD) (Gusella et al. 1983). Genes of smaller individual effect (oligogenes) have higher frequency and show complex patterns of inheritance. For polygenes the individual effects may be so small as to make such genes undetectable with current technology. Progress in identifying polygenes involved in complex inheritance may come through the development of arrays of single nucleotide polymorphisms (SNPs) in which there is a representation of the polymorphism in every human gene (see, for example, Chakravarti 1999).

Principles and Practice
Molecular Genetic Epidemiology – A Laboratory Perspective
Ian N.M. Day (Ed.)
© Springer-Verlag Berlin Heidelberg 2002

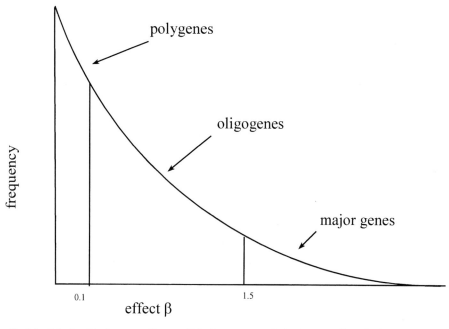

Fig. 1.1. Relationship between disease allele frequency and phenotypic effect (β) for three gene categories

1.2 Power and Sampling Considerations

For diseases with a Mendelian pattern of inheritance, power calculations giving information about error rates and sample size requirements can be relatively straightforward. For common diseases with a complex mode of inheritance this is no longer true. In this case examination of power usually involves estimating the sample sizes required to detect a gene that requires simplifying assumptions about the (unknown) underlying mode of inheritance.

There are two basic approaches to identification of the oligogenes involved in complex traits. The first is genetic *linkage*, which refers to the tendency of alleles at a disease and a marker locus to segregate together within a family if they are located physically close to each other on a chromosome. The second approach is *allelic association* (also called linkage disequilibrium), which is the tendency of a *specific* marker allele or alleles to be associated with a disease allele, which may be due to the close proximity of a disease and marker locus.

1.2.1 Linkage

For major genes, complex segregation analysis, which involves fitting a parametric model, has been used successfully to estimate genetic parameters such as gene frequency, dominance and penetrances (see, for example, Morton et al. 1991). Estimates from parametric models may be used in programs such as LINKAGE (Lathrop et al. 1984) to obtain a logarithm of odds (lod) score for linkage of markers to a disease gene. For complex traits, segregation analysis is faced with insurmountable difficulties. Alternatives that have been proposed include carrying out linkage analysis under several simple models in the hope that a single gene might have an effect large enough to be detected even under the incorrect model. However, "nonparametric" or, more correctly, weakly parametric approaches are usually employed to avoid fitting inappropriate multi-parameter models.

Lio and Morton (1997) evaluated both weakly parametric models and a parametric model, using the data presented by Davies et al. (1994), for insulin dependent diabetes (IDDM). In this study a sample of 95 sib pairs affected with IDDM were typed for 28 markers on chromosome 6. A number of non-parametric methods were used including the BETA (β) model (Morton 1996), Genehunter (Kruglyak et al. 1996) and MAPMAKER/SIBS (M/S) (Kryglyak et al. 1995). The COMDS program (Morton et al. 1991), which performs combined segregation and (single locus) linkage analysis, was used as a parametric model for comparison. The authors recognised that the data are very unsuitable for segregation analysis (with, for example, all normal sibs omitted). However, the results, in terms of the linkage evidence, are of some interest. Unlike the weakly parametric approaches the COMDS program is implemented only for single point analysis (single marker locus) and so is not able to obtain a multipoint lod score. Multipoint mapping is more powerful as it uses genotypic information from multiple markers in the map. Amongst the single point tests, however, COMDS gave the highest total lod and these lods were identical or higher to those obtained under the BETA model for markers closest to the three putative IDDM loci (IDDM 1, 5 and 8; Table 1.1).

The equivalence of combined segregation and linkage analysis to the weakly parametric approaches seems remarkable given the inadequacy of the ascertainment correction and the incomplete data consisting only of affected sib pairs with parents. However, it is unlikely that the equivalence would hold in all cases, motivating the use of the weakly parametric alternatives.

Risch (1990) developed the use of the ratio λ of the risk to relatives of a proband versus the population prevalence as an indication of mode of

Table 1.1. Single marker analysis of IDDM (lods)

Marker	Gene	COMDS	BETA	M/S	GENEHUNTER
D6S258	IDDM1	8.41	8.41	7.27	5.14
ESR	IDDM5	1.82	1.82	1.77	1.48
D6S264	IDDM8	1.02	0.59	0.50	0.35
Total (28 markers)	—	52.24	49.14	44.54	32.13

Table 1.2. Approximate minimum λ_S for 90% power to detect linkage

Number of affected sib pairs	λ_S
60	7
70	5
80	4
200	2

inheritance. The risk ratio λ_R for a type R relative of an affected individual is $\lambda_R = K_R/K$, where K is the population prevalence and K_R is the recurrence risk for a relative of an affected individual. For example, $\lambda_s = 3.0$ for a disease indicates that siblings of a proband have a three-fold increased risk of the same disease compared to the risk to individuals in the general population. As an example, a disease in which 10% of siblings of affected individuals are themselves affected with a disease for which the prevalence in the general population is 1/500, the value of λ_s is 0.1/0.002 = 50.0. The risk ratio decreases with the degree of relationship between the proband and relatives and the rate of decrease depends on the mode of inheritance.

For linkage analysis pairs of affected relatives with different degrees of relationship can be included using the identity by descent (IBD) between them. Identity by descent refers to the sharing of identical *ancestral* alleles between a pair of relatives (not just the sharing of an identical allele). The power to detect linkage depends on the risk ratio λ characterising the trait and also the recombination fraction (θ). When the risk to first-degree relatives is high (for example $\lambda_R > 5$), the recurrence risk pattern may be consistent with single locus inheritance. However, lower risk ratios suggest the presence of multiple loci influencing the phenotype. For sib pairs, the relationship between power and λ_S for a range of samples can be demonstrated. For a λ_S of 4, a sample size of approximately 80 affected sib pairs or greater is required to detect linkage with 90% power. Table 1.2 gives the approximate number of sib pairs required for a range of risk ratios.

This, however, is the *minimum* requirement as it assumes a fully informative marker and a recombination fraction between marker and disease gene of zero. Risch also examined the influence of increasing recombination and demonstrated that reduction in power was proportionally greater in small samples.

The question of nonindependence of sib pairs formed in larger sibships has received some attention. For families with multiple (s) affected or phenotyped siblings the number of possible sib pair comparisons is s(s − 1)/2. However, a sibship of this size contributes only s − 1 *independent* sib pairs, but all pairs are required for an efficient test. To compensate for this, Suarez and Hodge (1979) proposed to count all pairs but then divide each count by s/2 so that each pair carries weight as if there were only s − 1 pairs present. Wilson and

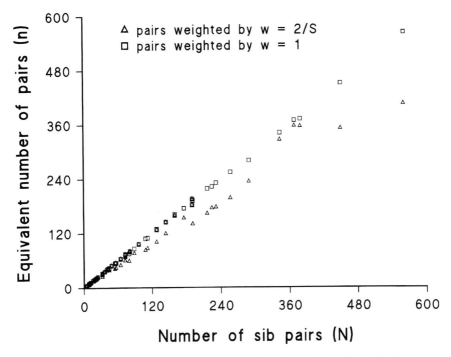

Fig. 1.2. The equivalent number of sib pairs compared to the actual number weighting pairs by 2/S where S is the number of siblings in the sibship and weighting all pairs equally (W = 1), for a range of samples with varying sibship sizes

Elston (1993), considering a quantitative trait, suggested that each sib pair be given equal weight to increase the power of the test, but the value of s – 1 be used to determine the number of degrees of freedom contributed by a sibship of size s. They did not show, however, that equal weights would increase power or give the predicted distribution. Collins and Morton (1995) examined the correlation (r) between identity by descent and a quantitative trait using data simulated under the null hypothesis of no linkage. They considered both equal weights (w = 1) and unequal weights for each sib pair (w = 2/S). The Fisher Z(r) transform was used allowing evaluation of the equivalent number of sib pairs (giving an indication of power) under both weighting schemes (Fig. 1.2). The figure shows that unequal weights per pair *lose* power as measured by the equivalent number of pairs. Therefore, for complex traits, the assumption of independence for all sibling pairs seems reasonable on the basis of available evidence.

The majority of complex phenotypic information is on a continuous or quantitative scale. Loci that influence a quantitative trait are termed QTLs. Examples of such quantitative phenotypes include various asthma measures, body mass index, and blood pressure. The power to detect QTLs depends on

Table 1.3. Number of sib pairs to detect a QTL under an additive model and a heritability of 10% assuming 80% power and using the top and bottom 10% of a quantitative phenotype

Allele frequencies	N pairs	Number of siblings to be screened to obtain the sample
0.10	1,647	19,120
0.30	1,482	17,357

the heritability (h^2), or the degree to which the trait is genetically determined, and the proportion of this heritability that can be attributed to an individual QTL. Approaches to quantitative traits are particularly sensitive to the sampling scheme and the utility of extreme discordant (very different) and extreme concordant (very similar) sib pairs has been emphasised. In particular, a sample comprising extreme discordant sib pairs from the tails of the quantitative distribution is known to be the most powerful strategy (Risch and Zhang 1995). Such pairs should share very few alleles identical by descent at the trait locus. With this sampling strategy a locus accounting for only 10% of the variance with an allele frequency of 0.3 can be detected with 1482 sib pairs (Table 1.3). Unlike concordant pairs the discordant sib pair approach is less susceptible to confounding due to other sources of sibling phenotype correlation (including shared environment). The practical difficulties of collecting such "ideal" samples are more of an issue, however. From Table 1.3 upwards of 17,000 siblings would have to be screened to obtain the desired number of discordant pairs fitting the ascertainment criteria. Sampling concordant pairs is less effective but such pairs are generally easier to obtain.

1.3 Models of Locus Action

Risch (1990) described both additive and multiplicative models of gene action. In the case of two loci if the genotypes at the first locus are G_i, $i = 1, n$, and at the second locus H_j, $j = 1, m$, then w_{ij} defines the penetrance of the two locus genotype $G_i H_j$. In a multiplicative or epistatic model the penetrance w_{ij} can be regarded as the product of two "penetrance factors" $w_{ij} = x_i y_j$, implying interaction between genotypes at the two loci. The relationship to familial risk is $\lambda_R = \lambda_{R1} \lambda_{R2}$ where 1 and 2 denote the two disease loci. Therefore, for a multiplicative model, the risk ratio λ_R is the product of "risk ratio factors" defined through penetrance factors for the two contributing loci.

In an additive model the penetrance w_{ij} can be regarded as a sum of two "penetrance summands" $w_{ij} = x_i + y_j$. This model may appear unrealistic implying for certain combinations of genotypes penetrances greater than

1. This can, however, be a good approximation to genetic heterogeneity where multiple loci contribute to a disease state but each locus is sufficient to cause disease on its own. In terms of familial risk, the total $\lambda_R - 1$ is a weighted sum of similar terms for each contributing locus. $\lambda_R - 1 = \lambda_{R1} - 1 + \lambda_{R2} - 1$.

If recurrence ratios of λ_R for a number of types of relationship are available the data can be used to estimate the number of loci involved in a complex disease and suggest whether the mode of action is additive or multiplicative. Risch (1990) performed an analysis of schizophrenia and concluded that the observed familial risks are consistent with a model of several loci acting epistatically. The best fit was a single major locus with $\lambda_1 = 3$ with a polygenic background of a large number of loci with small effects or two major loci each with $\lambda_1 = 2$ and a polygenic background. This finding gives some hope that one or two determinants of larger effect (contributing most to the overall familial association) might be detectable with current methods.

1.4 Maximum Likelihood Estimation

Typically, marker genotype data is unavailable on one or more sibs and IBD status cannot be fully determined. For this reason maximum likelihood methods are employed. However, where the IBD status is fully determined the method for obtaining the lod can be demonstrated using the data of Cox and Spielman (1989), which gives sharing proportions at HLA for sib pairs affected with type 1 diabetes (Table 1.4).

For Table 1.4 the allele sharing proportions for (0, 1 and 2 alleles IBD) are

$Z_0 = 10/137 = 0.073$

$Z_1 = 46/137 = 0.336$

$Z_2 = 81/137 = 0.591$

and the corresponding lod is therefore

$$\text{lod} = \log_{10} \frac{0.073^{10}\ 0.336^{46}\ 0.591^{81}}{0.25^{10}\ 0.5^{46}\ 0.25^{81}} = 16.978$$

Table 1.4. HLA sharing in sib pairs with type 1 diabetes (IDDM)

	Alleles IBD			
	0	1	2	N pairs
Observed	10	46	81	137
Expected	34	69	34	

Where IBD status is not fully determined, maximum likelihood sharing can be expressed as MLS $(Z_0, Z_1, Z_2) = \log 10$ $[L(Z)/L(0.25, 0.5, 0.25)]$, where $L(Z)$ are the maximum likelihood estimates of allele sharing proportions Z_0, Z_1 and Z_2. However, Holmans (1993) defined a "possible triangle" within which all biologically feasible models must fall and proposed a constrained model such that

$$Z_1 > 0,$$

$$Z_0 + Z_1 + Z_2 = 1,$$

$$Z_2 + Z_0 > Z_1,$$

and

$$Z_1 > Z_2$$

Under these constraints a maximum lod that simultaneously satisfies these restrictions is obtained. A further assumption (Kruglyak et al. 1995) of no "dominance variance" is equivalent to the constraint $Z_1 = 0.5$ giving a constrained single parameter model described by Z_2.

1.4.1 The β Model for Linkage

An alternative to these constrained models is the β model (Morton 1996), which describes the probability of 0, 1 or 2 alleles identical by descent in a pair of sibs by a single logistic parameter. Figure 1.3 shows the derivation of the probabilities in the β model. For a parent transmitting the same allele to a pair of affected sibs the probability is $1/(1 + e^\beta)$ and a different allele is $e^\beta/(1 + e^\beta)$. The β parameter is the log of the parent offspring recurrence risk λ_0. Following Risch (1990), the allele sharing probabilities for affected pairs in terms of recurrence risks under the β model are

$$Z_0 = 0.25/\lambda_S = 1\Big/\left(1 + e^{\beta f}\right)^2$$

$$Z_1 = 0.5 \cdot \lambda_0/\lambda_S = 2e^\beta\Big/\left(1 + e^{\beta f}\right)^2$$

$$Z_2 = 0.25 \cdot \lambda_m/\lambda_S = e^{2\beta}\Big/\left(1 + e^{\beta f}\right)^2$$

where λ_S is the recurrence risk in sibs, λ_0 the parent offspring and λ_m monozygous twins and where $f = 1$ for affected sib pairs. For a quantitative trait, f is replaced by the product $x_1 x_2$ where x_1 and x_2 are quantitative phenotypes defined as deviations from the trait population mean. As x is positive above the population mean and negative otherwise, f increases for phenotypically similar pairs in the tails of the distribution. This is in contrast to the conventional Haseman-Elston metric, which relates IBD sharing to the squared difference between traits $f = -(x_1 - x_2)^2$. In this formulation pairs with similar phenotypic values will be given equal weight wherever they are in the distribution.

Independent transmissions to pairs of affected sibs

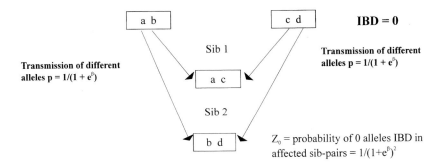

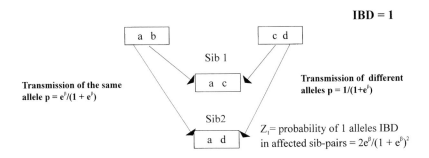

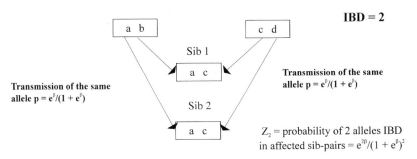

Fig. 1.3. Deviation of Z_0, Z_1, Z_2 allele sharing proportions under the β model for affected sib pairs

The single parameter β model has a number of advantages over multiparameter or constrained models. Transmissions from each parent are treated independently and the single parameter offers a significant increase in power (Collins et al. 1996; Lio and Morton 1997). Figure 1.4 shows the expected lod scores (ELODS) for single sib pairs from fully informative matings on both the

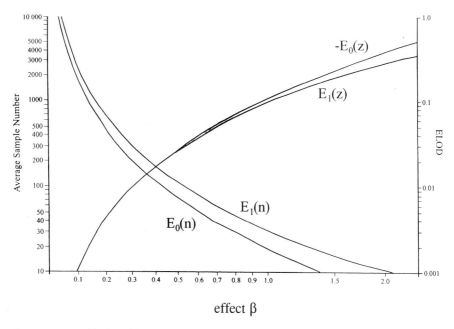

effect β

Fig. 1.4. Expected lods E_1 (Z), E_0 (Z) for single sib pairs for a range of β values and average sample sizes E_0 (n), E_1 (n) required to detect genes with a range of β values

null hypothesis of no linkage E_0 (Z) and the alternative hypothesis E_1 (Z) for various effects measured by the β parameter. Also shown are the average sample sizes (n sib pairs) required to detect or reject linkage under both hypotheses E_0 (n) and E_1 (n). It is clear that loci with values of β as small as 0.25 can be detected in a sample of a few hundred sib pairs. For genes of larger effect (for example, β = 1) 50 or fewer sib pairs are required.

Table 1.5 gives a comparison of multipoint lods from the β model and MAPMAKER/SIBS and GENEHUNTER using the IDDM data from Davies et al. (1994). The MAPMAKER/SIBS γ model fixes Z_1, the probability of 1 allele IBD, at 0.5 whilst the Δ model estimates two parameters and is less parsimonious. The data for three putative IDDM loci suggest increased power under the β model with a lod_1 of 11.28 at IDDM1, compared to 8.88 with MAP-MAKER/SIBS.

Lio and Morton (1997) also considered the relative efficiency of single locus and multipoint linkage tests and found (Table 1.5) that the efficiency of the single locus tests is only 67–75% of that of the full multipoint analysis, although this is unlikely to be a general conclusion for the comparison of single point and multipoint tests.

Wilkinson et al. (1998) applied the BETA program to a sample of 626 sib pairs with a quantitative asthma score. A lod_2 of 2.9 was obtained in a region

Table 1.5. Multipoint lods under alternative models

| Locus | β Model | | MAPMAKER/SIBS | | GENEHUNTER |
	Single locus efficiency	lod	γ lod	Δ lod	lod
IDDM1	0.75	11.28	8.88	11.61	9.73
IDDM5	0.73	2.48	2.42	2.48	2.32
IDDM8	0.67	0.96	0.90	1.03	0.76

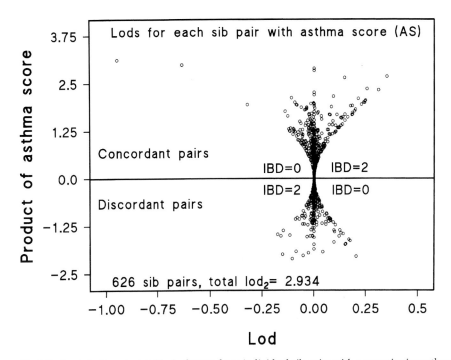

Fig. 1.5. Contributions to total lod of 2.934 from individual sib pairs with a quantitative asthma score

close to D12S97 on chromosome 12q. The individual contributions to the total lod from each sib pair are given in Fig. 1.5. Lods are plotted against the product of the asthma score $x_1 x_2$ for each pair. The figure illustrates both trait concordant ($x_1 x_2 > 0$) and discordant pairs ($x_1 x_2 < 0$) and the positive contributions to the lod of both types of pairs (concordant pairs where IBD = 2 and discordant pairs where IBD = 0). Three outliers are evident which reduce the total lod markedly and suggest heterogeneity.

1.5 Allelic Association

The dependence of allele frequencies at two loci is called allelic association or linkage disequilibrium. The term allelic association has been used in a more general sense to describe associations between nonsyntenic loci, resulting, for example, from population admixture.

Associations can be examined directly in candidate genes or within a candidate region. In the latter case, multiple-pairwise models have been devised to use information from a set of markers in a densely typed candidate region. These models depend on the expectation of greater association (when the effects of chance, mutation, and migration are negligible) as the disease locus is approached. Under this assumption, at least for major genes, a disease gene can in favourable cases be localised to within 50 kilobases (kb). Allelic association provides a means to map genes for disease susceptibility that is independent of linkage evidence and offers potentially greater resolution.

1.5.1 Family Based or Case Control

Data may be either family based or case control. Family-based procedures include the transmission disequilibrium test (TDT). The TDT (Spielman and Ewens 1996) and its many derivatives and extensions are motivated by the concern that sample-based (for example, case-control) methods are not robust to "spurious" associations arising from population stratification, migration and admixture. The TDT "considers parents who are heterozygous for an allele associated with disease and evaluates the frequency with which the allele or its alternate is transmitted to affected offspring". By considering only heterozygous parents the TDT is specific for association between *linked* loci and therefore protects against spurious associations. This favours a "trios" design with a single affected individual and genotyped parents. However, the test has now been extended to consider multiple siblings, quantitative traits (Allison 1997) and missing parents (Spielman and Ewens 1998) amongst others.

Morton and Collins (1998) have considered the relative efficiency of both TDT and case-control designs. The TDT can be parameterised in a way that assumes independent parental contributions and is therefore analogous to the β model for linkage, If ρ is allelic association then the probabilities of affection for a marker M are in the ratio $(1 - \rho)^2:(1 - \rho)(1 + \rho):(1 + \rho)^2$ for M'M', MM' and MM genotypes, where M' is any allele other than M. This ratio can also be expressed as $1:e^{\alpha}:e^{2\alpha}$ where $\alpha = \log[(1 + \rho)/(1 - \rho)]$. In terms of the TDT, the model is described in Table 1.6, where a is the number of times allele M is transmitted and b the number of times M' is transmitted to s affected children summed over all MM' parents.

In epidemiological studies affected cases are often compared to normal controls who are matched with respect to ethnic group and other attributes. Typ-

Table 1.6. The parameterised TDT test for M transmission from MM′ parents to an affected child

	Transmitted allele		Total
	M	M′	
Observed number	a	b	a + b
Expected frequency	$(1 + \rho)/2$	$(1 - \rho)/2$	1

Table 1.7. Some selective case-control strategies

Case	Control
1. Severely affected	Extremely normal
2. Affected with early onset	Normal, elderly
3. Positive family history	Negative family history
4. Affected	Normal with intense environmental exposure
5. Elderly survivor	Young
6. Affected, favourable covariate	Normal, unfavourable covariate

ically individuals are independent (unrelated). When discussing efficiency it is easy to compare the TDT design with a case-control study for the same number of individuals. A simple comparison can be made by examining the number of alleles available in the two designs. In the trio design N trios obviously comprise 3N individuals, therefore the equivalent number in the case-control design is 3N individuals with equal number of cases and controls. In the TDT a fully informative parent-child pair gives one transmission and one non-transmission, whereas two fully informative individuals in the sample-based case-control design contribute four alleles so the TDT efficiency is 1/2.

The possibility of using individuals from the extremes of the quantitative phenotype distribution favours the case-control design. For a phenotypic score selection of individuals in either tail of the distribution may be particularly advantageous. A series of case-control strategies that might be considered is shown in Table 1.7.

The extreme discordance strategy has been proposed for linkage (Risch and Zhang 1995) but is well suited to association. In this scheme (case 1) the extreme normals ("hypernormals") possessing a quantitative phenotype score well below the population mean may be compared with (severely) affected cases. Selection of families for early onset (case 2), which is efficient for rare major genes, has been useful in linkage studies. Case 3, which considers family history, is most applicable to genes with a moderate or large effect and is therefore less useful for common diseases. Case 4 has been used successfully to localise the CCR5 receptor for HIV-1 (Samson et al. 1996). Case 5 is relevant for genes involved in promoting longevity and case 6 is appropriate when an important predisposing factor is available as a covariate. An example is to contrast intrinsic asthma with atopic controls without asthma.

Under a multiplicative model, comparison of the efficiency of a case and hypernormal control design with TDT trios suggests that the efficiency of the latter is only 1/6 under the null hypothesis (Morton and Collins 1998). Such conclusions are of course model dependent. However, despite the low efficiency, multiplex sibships are suitable for analysis by the TDT, which also has the advantage of being useful in the study of imprinting (since the transmitting parents are identified). The TDT design is useful for sibships collected for linkage studies and remains valid if pedigrees are partitioned into nuclear families. Furthermore, the cost of studying a nuclear family and their S affected children may be less than the cost of studying S independent cases. However, diseases with late onset are unsuitable for a family-based design since the parents are often dead or unreachable. For this reason the sib-TDT has been proposed (Spielman and Ewens 1998). Unlike the TDT, the case-control design is not robust to spurious association although, at least for historically old and stable populations, the problem of inappropriate choice of controls has been exaggerated and the increased efficiency of such a design offers clear advantages.

1.6 The Malecot Model for Association

In addition to detecting association with polymorphisms in candidate genes directly, the relationship between allelic association and distance from a disease gene provides the potential for linkage disequilibrium mapping. Collins and Morton (1998) have implemented a model that parameterises both the decline of association with distance from a disease gene, the magnitude of association at the disease locus and the level of "spurious" association perhaps due to population admixture. This approach has been found to be quite powerful for major genes and in favourable cases can localise a disease gene to within 50 kb. This is higher resolution than can be achieved by linkage, for which mapping to within 1 cM ($\approx$1 Mb) is rarely achievable. For major genes the algorithm involves reducing (by stepwise merging) associated and nonassociated alleles at each marker in a candidate region into a 2×2 table from which the alleles with the largest χ_1^2 are selected stepwise. At least one allele is assumed to be associated for each marker in the candidate region whether χ_1^2 is significant or not. After correcting for sample enrichment (for example in a case-control haplotype study), the maximum likelihood coefficient of association is entered into a composite likelihood as $-\Sigma K_i(\hat{\rho}_i - \rho_i)^2/2$, where K_i is the information about the ith marker locus and ρ_i are the fitted values from the Malecot equation:

$$\rho_i = (1-L)M \cdot \exp(-\in \cdot d_i) + L$$

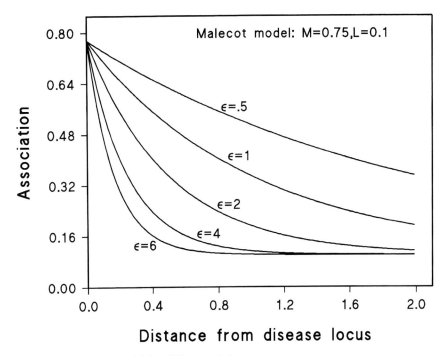

Fig. 1.6. The Malecot model for allelic association

The equation, which was developed to represent genetic isolation between populations as a function of geographical distance (Malecot 1948), is parameterised as M which approaches 1 for diseases with a monophyletic origin in which the disease mutation appears in a single haplotype transmitted from founders and is less than 1 otherwise. The parameter L represents the bias from assuming at least one associated allele at a marker and for spurious association due to population stratification. The number of generations and the disruption of haplotypes by recombination are represented by ϵ and the distance between the disease locus and the ith marker in the map is given by d_i. Figure 1.6 shows a representation of the Malecot model for M = 0.75 and L = 0.1. Alternative values for ϵ are illustrated; $\epsilon = 6$ implies that associations would only be detected within 400 kb, reflecting a long duration for founding disease haplotypes.

Lonjou et al. (1998) have evaluated this model for two major genes which have characteristics which make them potentially difficult for linkage disequilibrium mapping. Huntington's disease (HD) is such an example. HD was mapped to 4p16.3 by linkage with marker D4S10 (Gusella et al. 1983) but cloning was only achieved after haplotype analysis. Allelic association played no significant role in its localisation but it is an example of a highly polyphyletic gene with multiple HD mutations in large populations (MacDonald et

Table 1.8. Estimates of parameters under the Malecot model ($L = 0$). Mb Marker map in megabases; cM marker map in centimorgans; ε, M, Malecot parameters; S_D location under Malecot model; S_o known location of gene; Z_l lod for association

Estimate	HD		HFE	
	Mb map	cM map	Mb map	cM map
$\in$	1.04	0.68	·0.28	1.04
M	0.28	0.47	0.66	0.70
S_D	3.68	5.73	32.34	53.61
S_o	[3.64–3.80]	[4.09–4.38]	30.06	53.57
Z_l	47	45	296	307

al. 1992). Lonjou et al. (1998) collated all published HD data and examined the fit to the Malecot model. They also considered hemochromatosis (HFE), one of the most common recessive diseases in man. Linkage to the HLA region on 6p21.3 was demonstrated in 1976 although cloning was not achieved until an 8 Mb YAC contig was constructed and allelic association used to identify a 600 kb target later narrowed by haplotype analysis. One difficulty with the localisation of this gene is that of recombinational heterogeneity. There is a low level of recombination in this region which places HFE only 0.75 cM from HLA-A. However, physically, the distance is 4.6 Mb and the assumption that 1 cM ≈ 1 Mb gave a misleading indication to its location for many years. Results from the analysis using the Malecot model are shown in Table 1.8.

HD illustrates the typical case for which there is little recombinational heterogeneity in the candidate region. The physical map is more accurate than the genetic map and fits the model better (total lod for association $Z_l = 47$). The location $S_D = 3.68$ Mb is within the known bounds of the HD gene (3.64–3.80 Mb). The gene is highly polyphyletic, reflected in the low value for the M parameter (0.28). For HFE recombinational heterogeneity is so great that using the physical map distances gives a much poorer fit, suggested by the lower total lod. Furthermore, the error on location is very high at approximately 2.3 Mb compared with only 0.04 cM for the genetic map. This study shows, therefore, that for major genes the problems posed by both recombinational heterogeneity and polyphyletic origin can be overcome by this approach.

Methods for allelic association with quantitative traits are under development and are likely to be particularly important for complex traits. As for linkage, quantitative traits offer more power particularly where the definition of a qualitative phenotype depends on truncation of a quantitative variable. One approach being developed is to include information from sibships with parents in the form of a (weakly parameterised) quantitative TDT. Additional information can be included by treating the parents as a cohort sample. Association between marker loci and common disease alleles evaluated in this way can therefore be entered into the Malecot model as for major genes. Presumably, however, there will be considerably lower power than for major genes through a combination of the smaller individual effects and the inability to haplotype disease loci.

1.7 Candidate Genes for Allelic Association

By examining association between candidate gene alleles and disease, it is possible that disease alleles can be identified by direct association without the need for costly genome screens to identify candidate regions (and the subsequent difficult process of refining the candidate region). However, the limitations of candidate gene studies are many. One is the problem of disease complexity. Collins et al. (1999) conservatively noted 152 candidates for asthma and atopy and found a relatively poor correlation between the numbers of candidates in a region and the published linkage and association evidence in the 29 candidate regions identified. Most striking was chromosome 13, ranked a high 6 in the analysis, but having no clear candidates. At the opposite extreme was chromosome 1, ranked second to last at 28, with 19 functional candidates. The candidate gene approach is further limited by insufficient knowledge of the genome even for organisms for which the sequence is known. Amongst the organisms already sequenced the proportion of genes with *completely* unknown function is between 30% for *Mycoplasma genitalium* to around 58% for *Caenorhabditis elegans* (Collins 1999).

1.8 Significance Levels

A good deal of attention has been paid to defining an appropriate significance level or p value to accept a particular result as significant (for example, Lander and Kryglyak 1995; Morton 1998). For single gene disorders, a lod of 3 or more has continued to be accepted as a criterion for significance even though it was originally proposed for a sequential rather than fixed sample test (Morton 1995). In the case of a genome screen for a major locus, correction for multiple testing has been proposed by adding log10 of the number of markers tested to the lod criteria (Ott 1991). In this way a lod of at least $3.00 + \log 10 (300) = 5.48$ would be needed for acceptance as significant in a genome screen with 300 markers. However, such a procedure is too conservative as it treats all markers as independent, which will not be true unless they are so well spaced as to be effectively unlinked. The problem of dependency is further complicated because the strength of linkage increases when markers are closer together. Given these difficulties the lod 3 criteria has remained in general use for both single and multiple marker tests and experience has shown it to be a very robust criterion for major genes.

For complex traits there are greater difficulties. Lander and Kruglyak (1995) suggested that a lod of at least 2.2 be regarded as "suggestive", 3.6 "significant" and 5.4 "highly significant". These criteria are intended mainly to define significance for a single sample. More realistically, the initial identification of an interesting candidate region is likely to involve the acceptance of considerably lower lods, with confirmation or refutal through subsequent samples. Furthermore, for complex traits where single gene effects are relatively minor, it is most unlikely that any single sample would yield a lod of 5.4. The greater problem, therefore, is not that of defining criteria for acceptance of significance but in obtaining samples of sufficient size to detect a signal that will ultimately be confirmed. A possible solution is to perform meta-analysis.

1.9 Meta-Analysis

There has been a realisation that perhaps the greatest difficulty with mapping oligogenes is that of obtaining sufficiently large samples. Presumably, low power accounts for the lack of reproducibility of many studies. The main factor limiting power is more likely to be the sample size rather than the (perhaps relatively minor) increases provided by alternative analytical approaches. Even the largest study has relatively low power if the genes involved are of small effect and, even for genes of larger effect, a confirmatory sample is required. One solution to the sample size problem is through the combination of evidence, broadly called meta-analysis, which can be either through meta-analysis of published sources, *prospective* collaboration typified by national consortia like the CSGA (Marsh et al. 1997), or *retrospective* meta-analysis of independently published data sets. Collins et al. (1999) have performed a meta-analysis of 156 asthma and atopy publications to summarise available evidence, define (and rank) candidate regions and suggest candidate genes. One problem with the combination of evidence over studies is that different phenotypes, markers and maps are used in different studies. However, evidence from a single marker can be summarised by a nominal significance level p with $\chi_2^2 = -2 \ln(p)$ and a χ_1^2 for which the lod is $Z = \chi_1^2 / 2\ln(10)$. For studies reporting χ_m^2 a transformation through probability p to corresponding lod Z can be made (Morton 1998).

In other cases the published p value can be accepted but there is some concern that these are not always corrected for the number of tests (Bonferroni correction). One serious limitation of this type of meta-analysis is that of results reported only as "not significant" and also statistics such as the MLS (Kruglyak et al. 1995) for which the minimum value is zero. These can only be assigned $Z = 0$, which means that combination of evidence will be anticonservative. However, the value of this approach to meta-analysis is in the summary of results and the better definition of candidate regions prior to further study. Candidate regions were determined by merging samples falling within a bin of i ± 0.5 cM where i is an integer on the sex-averaged map. For each bin a summary lod Z, encompassing all published evidence, was determined and candidate regions defined by selecting Z in descending order but not within a 40 cM window.

In this way 29 candidate regions for atopy and asthma were defined. Graphical summaries for the two regions giving the highest lods (chromosome 6 and 5) are shown in Figs. 1.7 and 1.8. For chromosome 6 the HLA region is of considerable importance, most of the evidence coming from HLADRB1 and atopy (33 results), but the five asthma reports are also nominally significant and concordant (with a homogeneity χ^2 of 2.94) and a total lod of 5.42. Two reports on the PAFAH2 locus for platelet-activating factor which is distal to HLA suggest association of severe asthma with a low activity allele. For chromosome 5 the cytokine region has many reports and there appear to be different specificities. For example, ADRB2 is associated with asthma and bronchial reactivity whereas IL4 gives evidence of linkage and association to noncognate IgE (atopy).

Although useful, the many limitations of this form of meta-analysis (different phenotypes, different ascertainment schemes, choice of marker loci and inconsistent statistics) gives it no more than a summary role. Prospective collaboration requires each member to follow protocols relating to phenotype definition, sampling scheme and markers. Such uniformity is difficult and

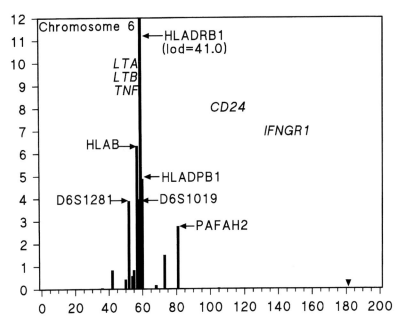

Fig. 1.7. Meta-analysis of asthma and atopy from published results for chromosome 6. Candidate genes in *italics*

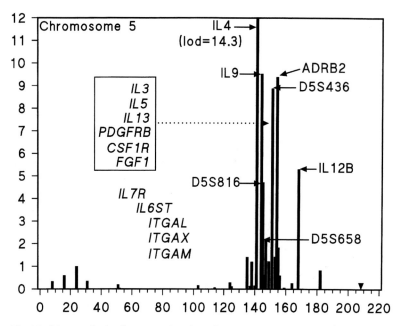

Fig. 1.8. Meta-analysis of atopy and asthma from published data for chromosome 5. Candidate genes in *italics*

expensive to achieve in practice. Therefore, *retrospective* collaboration is a viable alternative. Retrospective collaboration is typified by the international Consortium On Asthma Genetics (COAG; Lonjou et al. 1999). Each member of the consortium provides a data set that has already been published independently with no agreement about protocols. Coherence can be achieved using weakly parametric analysis (such as the β model) where the β parameter, information and location can be estimated for each study. Synthesis can be achieved through a lod table with the evidence evaluated at fixed points or by averaging estimates of location over studies weighted by their information. This approach has the further advantage of making combination of evidence from both linkage and association feasible.

1.10 Informatics

The area of informatics relevant to the identification of genes for common diseases is in the selection of useful polymorphisms, candidate genes, and maps of particular regions together with the analytical resources that are available.

1.10.1 Human Genome Resources

The process of disease gene identification is facilitated by the progress of the Human Genome Project. The current effort is focused on large-scale genomic sequencing. Furthermore, it is predicted that the emerging class of single nucleotide polymorphism (SNP) markers will soon dominate with 100,000 to be placed on the map by the year 2003 (Collins et al. 1998). However, much higher numbers are projected in the longer term since they are thought to be present at a density of 1 per 1 kb. Sequencing of the genome, for which a rough draft concentrating on gene-rich regions is predicted by 2001 and completion by 2005 or earlier, will provide detailed data on the entire catalogue of human genes. There is the possibility of chip-based allelic association in which SNPs in all human genes can be systematically tested. At present a relatively small number of single nucleotide polymorphic markers are available and a range of dinucleotide, tri and tetranucleotide markers are still widely used. The transcript map of expressed sequence tags (ESTs) is potentially useful for identifying transcripts expressed in certain tissues particularly if they map to an important candidate region.

Table 1.9 indicates some of the resources for mapping information that are available and gives an approximate classification of the material presented. The information is rather diverse and presented in similarly diverse ways. There is no single site that provides all that is likely to be required for a researcher interested in mapping genes for common diseases. The Genome DataBase (GDB) is the most general resource, although its funding status has been uncertain. It is particularly useful for tracing markers and genes through their synonyms, and providing primers and links to other sites. It offers relatively limited information on location. OMIM, or Online Mendelian Inheritance In Man, describes human genes and genetic disorders. It is an excellent resource for obtaining catalogue information on genes and links to important references. The Centre d'Etude du Polymorphism Humain (CEPH) provides a genotype database containing data submitted by more than 100 laboratories. Version 8.2 released in December 1998 contains genotypes for 11,995 genetic markers including more than 9,000 microsatellite markers for all human chromosomes. There are more than 2.5 million genotypes together with allele sizes and heterozygosity.

The Marshfield Medical Research Foundation (MMRF) provides maps and markers generated by the centre together with the data from the University of Utah and Genethon. The genetic maps of markers are comprehensive and contain more than 8,000 different loci. The Location DataBase (LDB) from the University of Southampton is concerned with the integration of maps of different data types (physical, radiation hybrid, genetic, cytogenetic, sequence-based and mouse homology). The maps include comprehensive linkage maps constructed from pairwise lod scores for both CEPH data and lods for genes collated by the GENATLAS database. The latter catalogues more than 9,000 genes and also provides maps, citations, and comparative maps and is

Table 1.9. Genome mapping resources on the WWW

	WWW site	Maps	Sequencing	Polymorphic markers	ESTs	Genes/Diseases
CEPH	http://www.cephb.fr/bio/			**		*
CHLC	http://www.chlc.org/	**		**		
GDB	http://www.hgmp.mrc.ac.uk/gdb/gdbhome.html	*		**	*	*
Genatlas	http://bisance.citi2.fr/GENATLAS/	*				**
Genbank	http://www.ncbi.nlm.nih.gov/web/search/index.html		**		*	**
Genecards	http://bioinformatics.weizmann.ac.il/cards/					**
Genemap 98	http://www.ncbi.nlm.nih.gov/genemap/	**			**	*
Genethon	http://www.genethon.fr/	**		**		
HGBASE	http://hgbase.interactiva.de/			**		
LDB	http://cedar.genetics.soton.ac.uk/public_html/	**		*	*	*
Marshfield	http://www.marshmed.org/genetics/	**		**		
MGD	http://www.informatics.jax.org/	*				
Nomenclature	http://www.gene.ucl.ac.uk/nomenclature/					*
OMIM	http://www.hgmp.mrc.ac.uk/omim/					**
Sanger	http://www.sanger.ac.uk/	*	**			
UDB	http://bioinformatics.weizmann.ac.il/udb	**		*	*	
Unigene	http://www.ncbi.nlm.nih.gov/UniGene/index.html				**	
Whitehead	http://www.genome.wi.mit.edu/	*		*	**	*

*Useful; **good

particularly useful for disease genes and phenotypes. The Unified DataBase (UDB) from the Weizmann Institute is also concerned with the integration of mapping information and presents the locations on an approximate megabase scale. Also from the Weizmann Institute is the GENECARDS database which catalogues human genes, their products and their involvement in disease. The information presented is automatically updated from external databases.

There are a number of sites concerned with physical mapping of the human genome. The target of the human genome mapping project has been to construct physical maps firstly from yeast artificial chromosome (YAC) contigs, followed by cosmid contigs and sequence-ready contigs. Radiation hybrid mapping has been useful to localise monomorphic segments such as expressed sequence tags (ESTs). The Whitehead Institute map provides information about locations for 24,000 sequence tagged sites, many derived from ESTs together with primer sequences, PCR conditions and other information. There is also a radiation hybrid map of over 14,000 STSs. GENEMAP98 is presented by the international RH mapping consortium. The map provides locations for ESTs derived from perhaps more than 30,000 genes mapped in radiation hybrids. The Sanger Centre provides sequence information for a number of genome regions and is concentrating in particular on chromosomes 1, 6, 9, 10, 13, 20, 22 and the X and is one of a number of major sequencing centres worldwide. HGBASE (for *Human Gene Bi-Allelic Sequences*) is a database of intragenic sequence polymorphisms. It is primarily designed for association studies based on SNPs, but covers all types of intragenic variation.

1.10.2 Analytical Resources

Software for the analysis of data for complex traits consists mainly of programs written by research groups to implement specific models and approaches. An important repository of software for the analysis of complex traits is maintained at: http://linkage.rockefeller.edu/soft/. Some of the many packages available are briefly described here. The BETA program for linkage analysis of sib pairs, ALLASS for allelic association under the Malecot model and COMDS for combined segregation and linkage analysis have already been described. There are a number of other programs for sib pair analysis which include MAPMAKER/SIBS and also ASPEX (*Affected Sib Pairs EXclusions* mapping) for multipoint exclusion mapping allowing both sex-specific marker maps and considering maternal and paternal transmissions separately. The GENE-HUNTER program is more useful for extended pedigrees and includes both parametric and nonparametric analysis with many additional features such as haplotype reconstruction and graphical output. It also features the TDT and variance component approaches. Also concerned with variance component mapping of quantitative trait loci is SOLAR (Sequential Oligogenic Linkage Analysis Routines) which is capable of multipoint variance component linkage analysis in pedigrees of arbitrary size and complexity. SOLAR can be used to model many complex genetic and environmental interactions and the approach holds some promise (see, for example, Williams and Blangero 1999) although it is not clear that selected (nonrandom) samples can be accommodated. The MIM program (Multipoint Identical by descent Method) is also concerned with partitioning the genetic variance of quantitative traits using nuclear families.

1.11 Summary

For Mendelian diseases the strategy for designing studies usually involves confirming that the sample collected is sufficient to detect linkage with reasonable power. In complex traits the underlying genetic model for the disease is virtually always completely unknown making reliable power studies difficult and dependent on simplifying assumptions. However, useful information about sample sizes in sib-pair studies can be obtained. The utility of linkage disequilibrium mapping for common diseases has yet to be fully demonstrated although SNP-based arrays representing polymorphisms in all human genes must hold great promise for the future. Presently meta-analysis, most practically in the form of retrospective studies, offers the potential to increase power through combination of samples and therefore provide more reliable and reproducible evidence for common disease genes.

References

Allison DB (1997) Transmission-disequilibrium tests for quantitative traits. Am J Hum Genet 60:676–690
Chakravarti A (1999) Population genetics – making sense out of sequence. Nat Genet Suppl 21:56–60
Collins A (1999) Mapping in the sequencing era. Hum Hered 50:76–84
Collins A, Morton NE (1995) Nonparametric tests for linkage with dependent sib-pairs. Hum Hered 45:311–318
Collins A, MacLean CJ, Morton NE (1996) Trials of the β model for complex inheritance. Proc Natl Acad Sci USA 93:9177–9181
Collins A, Ennis S, Tapper W, Morton NE (2000) Mapping oligogenes for atopy and asthma by meta-analysis. Genet Mol Biol 23:1–10
Collins A, Morton NE (1998) Mapping a disease locus by allelic association. Proc Natl Acad Sci USA 95:1741–1745
Collins FS, Patrinos A, Jordan E, Chakravarti A, Gesteland R, Walters L, Fearon E, Hartwelt L, Langley CH, Mathies RA, Olson M, Pawson AJ, Pollard T, Williamson A, Wold B, Buetow K, Branscomb E, Capecchi M, Church G, Garner J, Gibbs RA, Hawkins T, Hodgson K, Knotek M, Meisler M, Rubin GM, Smith LM, Smith RF, Westerfield M, Clayton EW, Fisher NL, Lerman CE, McInerney JD, Nebo W, Press N, Valle D (1998) New goals for the US Human Genome Project. Science 282:682–689
Cox NJ, Spielman R (1989) The insulin gene and susceptibility to IDDM. Genet Epidemiol 6:65–69

Davies JL, Kamaguchi Y, Bennett ST, Copeman JB, Cordell HJ, Pritchard LE, Reed PW, Gough SCL, Jenkins SC, Palmer SM, Balfour KM, Rowe BR, Farrall M, Barnett AH, Bain SC, Todd JA (1994) A genome-wide search for human type-1 diabetes susceptibility genes. Nature 371:130–136

Gusella JF, Wexler NS, Conneally PM, Naylor SL, Anderson MA, Tanzi RE, Watkins PC, Ottina K, Wallace MR, Sakaguchi AY, Young AB, Shoulson I, Bonilla E, Martin JB (1983) A polymorphic DNA marker genetically linked to Huntington's disease. Nature 306:234–238

Holmans P (1993) Asymptotic properties of affected sib-pair linkage analysis. Am J Hum Genet 12:362–374

Kerem BS, Rommens JM, Buchanan JA, Markiewicz D, Cox TK, Chakravarti A, Buchwald M, Tsui L-C (1989) Identification of the cystic fibrosis gene – genetic analysis. Science 245:1073–1080

Kruglyak L, Daly MJ, Lander ES (1995) Rapid multipoint linkage analysis of recessive traits in nuclear families, including homozygosity mapping. Am J Hum Genet 56:519–527

Kruglyak L, Daly MJ, Reeve-Daly MP, Lander ES (1996) Parametric and non-parametric linkage analysis: a unified approach. Am J Hum Genet 58:1347–1363

Lander ES, Kryglyak L (1995) Genetic dissection of complex traits. Guidelines for interpreting and reporting linkage results. Nat Genet 11:241

Lathrop GM, Lalouel JM, Julier C, Ott J (1984) Strategies for multilocus linkage analysis in humans. Proc Natl Acad Sci USA 81:3443–3446

Lio P, Morton NE (1997) Comparison of parametric and nonparametric methods to map oligogenes by linkage. Proc Natl Acad Sci USA 94:5344–5348

Lonjou C, Collins A, Ajioka RS, Jorde LB, Kushner JP, Morton NE (1998) Allelic association under map error and recombinational heterogeneity: a tale of two sites. Proc Natl Acad Sci USA 95:11366–11370

Lonjou C, Collins A, Ennis S, Tapper W, Morton NE (1999) Meta-analysis and retrospective collaboration: two methods to map oligogenes for atopy and asthma. Clin Exp Allergy 29 [suppl 4]:57–59

MacDonald ME, Novelletto A, Lin C, Tagle D, Barnes G, Bates G, Taylor S, Allitto B, Altherr M, Myers R, Lehrach J, Collins FS, Wasmuth JJ, Frontali M, Gusella JF (1992) The Huntingtons-disease candidate region exhibits many different haplotypes. Nat Genet 1:99–103

Malecot G (1948) Les mathématiques de l'heredité. Maison and Cie, Paris

Marsh DG, Maestri NE, Freidhoff LR, Barnes KC, Togias A, Ehrlich E, Beaty T, Duffy D, Rosenthal R, Imani F, Dunston G, Furbert-Harris P, Malveaux F, Ober C, Cox NJ, Lester LA, Peterson R, Gidley H, Pluzhnikov A, Anderson J, Solway J, Leff A, Wolf R, Wylam M, Kurtz B, Richardson A, Parry R, Blumenthal MN, King RA, Oetting W, Drury D, Rosenberg A, Daniels L, McEvor C, Lou J, Hamra M, Brott M, Rich SS, Spray BJ, Weber JL, Yuan B, Wang ZY, Bleecker ER, Amelung P, Rechstiner B, Meyers DA, Samet J, Wiesch D, Xu JF, Murphy S, Banks-Schlegel S (1997) A genome-wide search for asthma susceptibility loci in ethnically diverse populations. Nat Genet 15:389–392

Morton NE (1995) Sequential tests for the detection of linkage. Am J Hum Genet 7:277–318

Morton NE (1996) Logarithm of odds (lods) for linkage in complex inheritance. Proc Natl Acad Sci USA 93:3471–3476

Morton NE (1998) Significance levels in complex inheritance. Am J Hum Genet 62:690–697

Morton NE, Collins A (1998) Tests and estimates of allelic association in complex inheritance. Proc Natl Acad Sci USA 95:11389–11393

Morton NE, Shields DC, Collins A (1991) Genetic epidemiology of complex phenotypes. Ann Hum Genet 55:301–314

Ott J (1991) Analysis of human genetic linkage. John Hopkins University Press, Baltimore

Risch N (1990) Linkage strategies for genetically complex traits. 1. Multilocus models. Am J Hum Genet 46:222–228

Risch N, Zhang HP (1995) Extreme discordant sib-pairs for mapping quantitative trait loci in humans. Science 268:1584–1589

Samson M, Libert F, Doranz BJ, Rucker J, Liesnard C, Farber CM, Saragosti S, Lapoumeroulie C, Cognaux T, Forceille C, Muyldermans G, Verhofstede C, Burtonboy G, Georges M, Imai T, Rana S, Yi YJ, Smyth RJ, Collman RG, Doms RW, Vassart G, Partmentier M (1996) Resistance to

HIV-1 infection in Caucasian individuals bearing mutant alleles of the CCR-5 chemokine receptor gene. Nature 382:722–725

Spielman RS, Ewens WJ (1996) The TDT and other family-based tests for linkage disequilibrium and association. Am J Hum Genet 59:983–989

Spielman RS, Ewens WJ (1998) A sibship test for linkage in the presence of association: the sib transmission-disequilibrium test. Am J Hum Genet 62:450–458

Suarez BH, Hodge SE (1979) A simple method to detect linkage for rare recessive diseases: an application to juvenile diabetes. Clin Genet 15:126–136

Wilkinson J, Grimley S, Collins A, Thomas NS, Holgate ST, Morton NE (1998) Linkage of asthma to markers on chromosome 12 in a sample of 240 families using quantitative phenotype scores. Genomics 53:251–259

Williams JT, Blangero J (1999) Comparison of variance components with sib pair based approaches to quantitative trait linkage analysis in unselected samples. Genet Epidemiol 16:113–134

Wilson AF, Elston RC (1993) Statistical validity of the Haseman-Elston sib-pair test in small samples. Genet Epidemiol 10:593–598

2 Human DNA Sampling and Banking

2.1 Introduction

The first era of DNA research, from the discovery of DNA itself by Friedrich Miescher in 1869 (Harbers 1969) to the completion of the human genome sequencing project scheduled for the year 2005 (Table 2.1), is about to end. The argument put forward in this chapter is that the existing DNA sampling and storing techniques were set up to satisfy different goals and needs than those that will be required in the forthcoming era of genomics and of large-scale population genetics. The human genome diversity project aims at the definition and explanation of every polymorphism, every rare or sporadic germ-line or somatic mutation, every Mendelian, polygenic or complex disease and every inherited phenotypic trait on a global scale (Cavalli-Sforza et al. 1991; Kidd et al. 1993; Cavalli-Sforza 1998; Weiss 1998). Such projects may seem immense today but so did the various genome projects a few years ago. Research for high-throughput analytical methodologies is producing most encouraging results. There are already analytical techniques for genotyping several thousands of individuals per technician per day (Day et al. 1998) and DNA extraction is becoming a bottleneck of progress.

2.2 A Brief History of DNA Research

In the past 130 years, we have seen four major DNA-related disciplines arise. The first studies were concentrated on the chemical and physicochemical properties of the genetic material (Hotchkiss 1979, 1995). The model organisms were, naturally, the simplest available, i.e. bacteriophages and their hosts. As for all chemistry and physical chemistry, samples had to be adequately concentrated and of the highest possible purity. Yield and purity were, therefore, the primary criteria of evaluation of DNA extraction methods at those initial steps.

With a successful model of the primary and secondary structure of DNA (Watson and Crick 1974), research entered a second phase of development.

Principles and Practice
Molecular Genetic Epidemiology – A Laboratory Perspective
Ian N.M. Day (Ed.)
© Springer-Verlag Berlin Heidelberg 2002

Table 2.1. Landmarks in DNA research

Year	Development	Developer/Reference
1869	Discovery of DNA	F. Miescher (Harbers 1969)
1944	DNA is the genetic material	O.T. Avery, C.M. MacLeod and M. McCarty (Hotchkiss 1979)
1953	The double helix model	J.D. Watson and F.H.C. Crick (Watson and Crick 1974)
1965	The genetic code	M. Nirenberg, H.G. Khorana and R.W. Holley (Anonymous 1968)
1968	Chemical synthesis of oligonucleotides	S.A. Narang et al. (Narang et al. 1968)
1973	DNA electrophoresis on agarose gels	P.A. Sharp et al. (Sharp et al. 1973)
1975	The Southern blot	E.M. Southern (Southern 1975)
1977	DNA sequencing	F. Sanger et al. (Sanger et al. 1977), A.M. Maxam and W. Gilbert (Maxam and Gilbert 1977)
1982	GenBank established	
1983	Automated DNA sequencing	L. Hood et al. (Hunkapiller et al. 1983; Hood et al. 1987)
1986	PCR	K. Mullis et al. (Mullis et al. 1986)
1987	Human Genome Organisation	
1989	H. influenzae genome sequenced	
1990	Human Genome Sequencing Project	
1991	Human Genome Diversity Project	L.L. Cavalli-Sforza et al. (Cavalli-Sforza et al. 1991)
1999	C. elegans genome sequenced	(Wilson 1999)
2001	A map of ~150,000 human SNPs	(Woodman 1999)
2005	H. sapiens genome completed	(Wadman 1999)

DNA metabolism and the nature of mutation became the central themes. The simplest available chromosomes (including plasmids) continued to provide the primary material. DNA became the substrate of experimental enzymatic reactions, and, therefore, preparations still had to be of maximum purity and integrity. The enzymes of DNA anabolism, catabolism, recombination and repair (Gussin et al. 1966; Zimmerman et al. 1967; Gellert et al. 1968; Melgar and Goldthwait 1968a,b; Baltimore 1970; Temin and Mizutani 1970; Kornberg et al. 1989), the deciphering of the genetic code (Khorana 1965; Nirenberg 1965; Anonymous 1968), DNA labelling techniques (Weiss and Richardson 1966) and the chemical synthesis of oligonucleotides (Narang et al. 1968) were all products of the intense DNA biochemistry of the 1950s and 1960s. In the 1970s, DNA biochemistry grew into a revolutionary recombinant-DNA technology that allowed cloning, sequencing and banking of isolated genome segments (probes).

The combination of DNA electrophoresis on agarose gels (Sharp et al. 1973) with the probing techniques of E. Southern (Southern 1975) signalled the beginning of what we now strictly call molecular genetics, i.e. the study of genetic variation between individuals at the level of DNA. However, in the first two decades of molecular genetics, the bulk of the work was aimed at an understanding of gene structure and function in higher organisms (exons, introns, promoters and other regulatory elements). The commonest theme became gene expression. The dominant methodologies were those of cellular and classical organismic genetics involving one or few specimens. Development of preparative and analytical methods was by and large for the study of gene products (RNA and protein) in selected or engineered cell lines, rather than for the study of DNA itself. The segments of interest (mostly single-copy genes) were only tiny fractions of the extracts (entire complex genomes) and both genomic and probe DNA preparations routinely involved enzymes for restriction and labelling. The most impressive technological advances of the early 1980s, like the automation of DNA sequencing (Hunkapiller et al. 1983; Hood et al. 1987) and the invention of the polymerase chain reaction (PCR, Mullis et al. 1986), were still in the biochemical field. Good yields of pure, high molecular weight DNA continued, therefore, to be pivotal to these developments.

Oligonucleotide synthetic chemistry and the rapid development of PCR techniques triggered a second revolution in molecular genetics. PCR produces designed probes and/or templates in such amounts and in such simple chemical media that further purification is not usually required. Most post-PCR analytical methods perform well, or can be adapted to do so, with crude PCR products. What is more, the so-called *amplification* of DNA by PCR is itself largely insensitive to impurities or degradation of DNA substrates. Therefore, for PCR-mediated analysis, high yield, purity and integrity of the primary genomic DNA ceased to be as important as they were previously.

However, due I believe to the long tradition, developers of DNA extraction methods continue to pay excessive attention to the quantity, purity and integrity of their product but have so far brought little improvement with regard to throughput components, i.e. speed, ease, safety and cost-effectiveness of extraction.

2.3 Goals and Needs of Human Population Genetics

The definition of the human genetic diversity project does not limit research to fighting disease and improving health. However, these goals have been and will continue to be the first priority of human population genetics in the foreseeable future. Health-related population genetics is known as genetic epidemiology (Schull and Weiss 1980). The common aim is to examine associations (hypothesised causal relations) between genes and phenotypic traits, or between gene alleles and disease traits. The basic methodological designs (Table 2.2) are inspired from classical epidemiology and classical genetics (Andrieu and Goldstein 1998).

Table 2.2. Common sample requirements for human genetics

Study design	Typical sample size	Time scale	DNA extraction criteria
Case-control	10–1000	Point	Cost/speed/yield
Retrospective	<100	Point/regular	Yield/stability
Cohort	100–10,000	Point	Throughput/stability
Follow-up	100–1000	Period in the past	Stability/special resources/yield
Prospective	100–1000	Point	Stability/throughput
Survey	>10,000	Period	Cost/speed/stability
Genetic screening	>10,000	Regular sampling	Cost/speed/yield
Neonatal screening	>10,000	Regular sampling	Cost/speed/yield
Pharmacogenetics/ prevalence studies	>1000	Point	Throughput
Large pedigrees	20–100	Point	Yield/stability
Nuclear families	10–1000	Period	Yield/stability
Mendelian inheritance	<10–100	Point	No special requirements
Non-Mendelian inheritance	100–1000	Point	Yield/stability/throughput

Case-control studies compare frequencies of candidate-gene alleles in two groups: individuals that present the disease trait against those that do not (Khoury and Beaty 1994). The smaller the penetrance of the genetic factor and/or the larger the number of genetic factors involved, the larger the recruited groups need to be. Case-control studies require, however, a single sampling operation that is usually completed within a limited period (a single point) of time. Therefore, the cost, speed, and, eventually, the yield (for genome scans) of the DNA extraction method to be used are more important than the purity and the long-term stability of the extracts.

In a subset of case-control studies (*retrospective studies*) inferences about an exposure to putative causal factors derive from data relating to characteristics of persons under study or to events or experiences in their past. The availability of such data may limit the size of the recruited groups and may render the samples unique (Scheibe 1981). The yield and the long-term stability of the sampled material are most important for prolonged exploitation of unique samples.

Cohorts are defined populations which, as a whole, are followed in an attempt to determine distinguishing subgroup characteristics. The size of a cohort depends on the anticipated size of the smallest subgroup to be identified. In *longitudinal studies*, variables relating to cohort subgroups are measured over a relatively long period of time (several years) in order to assess the outcome of exposures, procedures or effects of a characteristic, e.g. occurrence of disease (*follow-up studies*). In *prospective studies*, a cohort is observed for a sufficient number of years to generate incidence or mortality rates after the selection of the study group. In longitudinal designs, sampling and analysis are often remote in time. Stability of the sample is, in that case, important. Frequently, in the context of a longitudinal study, geneticists rely on sub-optimal archival biological material collected for other diagnostic purposes and on special extraction methods to produce sufficient DNA for analysis. These difficulties limit the sample size and may justify costly DNA-extraction options.

Genetic epidemiological designs also intersect with *health surveys*, which are systematic collections of factual data pertaining to health and disease in a human population within a given geographical area. These may take the form of an organised procedure performed on large groups of people for detecting disease (*mass screening*). Mass screening may include genetic tests (Bacon and Schilsky 1999) but *genetic screening* usually refers to genotyping (or karyotyping), at a regular rather than periodic basis, of persons already associated with disease or predisposition to disease, or of persons who may transmit disease to descendants (Grann et al. 1999). Genetic screening includes *prenatal genetic screening* whereas *neonatal screening* for diagnosis, prognosis, monitoring or counselling may include genetic tests (Bradley et al. 1993). The limits of such massive sampling operations are usually financial. The cost of DNA extraction in consumables and salaries is the first factor to consider. The speed of extraction becomes particularly important in surveys that are to be completed within a defined period of time, or in screenings, when large numbers of samples are received daily. For cost-effectiveness, it is desirable that large collections are used in multiple studies beyond the initial prospects, but high-yield/purity methods as well as storage may disproportionally raise the costs.

Further designs originate from the classical genetic concept of *linkage* (Elston 1998a). Genetic markers are highly polymorphic loci that are not causally associated with the phenotypic trait of interest but do co-segregate with it. Co-segregation is extremely low frequency, or absence, of recombination events between the causal locus and the linked marker due to the physical proximity of the two loci on a chromosome. Linkage studies may employ large pedigrees, a set of small families, collections of *affected sib pairs* (Yang et al. 1997; Sun et al. 1998), "trio" collections (proband and parents; Flanders and Khoury 1996) or other case-control schemes. A common aim of such studies is the identification and mapping of novel genes.

The scale of linkage studies regarding sample size sounds much smaller than that of the epidemiological designs above. This is true for single fully penetrant loci with Mendelian inheritance, even when many families and many markers have to be analysed. In fact, family studies and population association studies, which may take dimensions of a mass screening, are the two extremes of a continuum (Houwen et al. 1994). In genetic analysis, power is generated either by studying many loci in one specimen (a high DNA yield needed), or one locus in many specimens (high throughput needed). Much of current human genetics is concerned with genes with reduced penetrance or oligogenic, multifactorial (polygenic) or complex disorders (myocardial infarction, diabetes, essential hypertension, cancers, osteoporosis, multiple sclerosis, ankylosing spondylitis, to mention a few; Khoury and Yang 1998). For such disorders, and for quantitative traits in general, study designs involve both, families and large population groups (Elston 1998b). Accordingly, a sampling strategy may combine solutions with regard to source-tissue, DNA-extraction method and/or storage.

New, "preventive" disciplines are exclusively concerned with healthy populations and/or with genes that are only indirectly related to health. To determine the prevalence of resistance to a pathogen (Lucotte and Mercier 1998), one should not wait until an entire population is infected. Similarly, to decide who is susceptible to what environmental hazards, genetic risk assessments are made on yet unaffected individuals (Merikangas 1993; Newman et al. 1997; London et al. 1999). Pharmacogenetics (Linder and Valdes 1999) is a young discipline that will tailor drug treatment to fit the genetic background of individual patients. Beside a disease gene, subjects of pharmacogenetic studies are genes involved in drug metabolism and tolerance. Knowledge of the relevant genotypic distributions of healthy populations at early stages will greatly assist rational drug design.

Within the framework of the human diversity project, we may anticipate that complex trait genetics will expand far beyond the health domain as soon as ethical/legal, methodological, technical and financial issues are resolved. Research has already begun on questions of origin, history and evolution of human populations (Cavalli-Sforza 1998). Despite history, behavioural and social genetics may be reborn on new ethical, legislative and methodological bases. It is likely, for example, that in the next few decades entire populations of young people will be screened for a panel of vocational genes before they choose their career.

It is clear from the above that contemporary genetic studies require many more DNA samples than ever before. A family of "National Health and Nutrition Examination Surveys" (NHANES; undertaken by the Centers for Disease Control and Prevention; USA; Steinberg et al. 1997) is, perhaps, one of the largest collections of human specimens amenable to genetic analysis reported to date. This includes mononuclear cell samples from 19,553 individuals collected between 1966 and 1997. A prospective genetic study of pregnancy and foetal growth (I.N.M. Day, A.C. Collins, S. Ye, N. Morton, E. Spanakis and S. O'Dell, unpubl.) requires an initial collection of 20,000 DNA samples from a local female population. Subsequent family-based association studies of several thousand pregnancies will require additional samples from the mother-to-be, father, offspring and, eventually, the mother's parents for each pregnancy. The time scale of this project, including the analysis, is 5 years.

There is no doubt that more ambitious studies will soon become routine even in small laboratories. The odds are that DNA sampling and banking demands will continue to increase. A large proportion of those samples will come from people who may consent to participate in a population study but, as they do not directly benefit from it, they may not have time to spare. Population geneticists will have to move from the hospital to places where masses congregate: educational institutions, work places, shopping centres, entertainment venues, airports etc., and to adopt marketing-like methods to reach their target groups. Such studies will probably involve complex designs with large cohorts, or surveys, and subset groups or families for more detailed retrospective and/or prospective study of many loci. Large population samples constitute invaluable mines of genetic information and should be stored safely for as long as possible. How would the current sampling technologies perform in the above situations?

2.4 Selection of the Source Tissue

In principle, any type of nucleated cells can serve as a DNA source. The primary criteria for selecting the source tissue are not only the amount and relevance of information each tissue can provide but also the cost and the feasibility of sampling. For most studies, the options are severely restricted and, for many studies, there is no option but an already existing unique resource (Table 2.3).

Table 2.3. Common DNA sources

Biological material	Advantages	Disadvantages	Main uses
Blood	Rich in phenotypic information; large yield	Complex DNA extraction; biohazard	Small-scale/ retrospective studies
Solid tissues	Histopathological information; may contain somatic mutations	Complex collection; complex DNA extraction; biohazard	Somatic genetics; cytogenetics; small-animal and plant genetics
Cell lines	Immortal; unlimited yield; easy DNA extraction	Evolution in vitro; biohazard; difficult establishment	Cytogenetics; retrospective studies
Rare specimens	Unique	Very low yield; complex DNA extraction	Forensics; paleontology; archaeology; studies of archival material
Mouth mucosa	Easy collection and handling	Frequently contaminated with foreign DNA	Large-scale human population genetics

2.4.1 Blood

For example, 10 ml of normal blood yield about 100–400 µg of pure, high molecular weight DNA. Carefully managed, this amount is sufficient for several tens of thousands of PCRs and, therefore, can supply molecular information for at least the same number of loci, equivalent to several high-resolution whole genome scans. Since only one high-resolution whole genome scan will theoretically ever be required per individual, the remaining DNA can serve to genotype an equivalent number of "candidate" loci. In fact, microarray technology (Brown and Botstein 1999) moves toward the possibility of re-sequencing the entire genome from a single 10-ml blood specimen.

In addition, blood is a rich source of phenotypic information for all measurable circulating peptides and metabolites (van Aken 1986). Many existing blood collections are only by-products of biochemical screening. Before the invention of PCRs, the amount of DNA required for molecular analysis would alone justify blood collection. In the post-PCR years, the amount of genetic information a blood sample can provide is excessive (Angelico and Del Ben 1992). The design of sampling and the recorded phenotypic and non-genetic (demographic, environmental) data that accompany the specimens limit the number of biological questions the sample can really answer. Thus, a sample may be re-visited only for investigations that are related to the original purpose of sampling and at a rate much slower than what its collectors would have hoped.

Blood sampling requires qualified venesectors, special sterile equipment (needles and syringes), chemicals (disinfectant, anti-coagulant), and various accessories. For large-scale studies, these may add up to a considerable, if not prohibitive, cost. Classified as a biohazard and being an excellent substrate for air-borne contaminants, blood requires special handling at every step before DNA extraction. Edetic acid (EDTA) and heparin (frequently used as an anti-coagulant) as well as haem, the red group of haemoglobin, may inhibit PCR and/or interfere with DNA (Hall and Axelrod 1977; Gustafson et al. 1987; Yokota et al. 1999) and their removal may complicate purification.

For all these reasons, blood is now less attractive as a DNA source. De novo large-scale blood sampling is no longer justifiable unless accompanied by biochemical measurements or done at places where blood is readily available (hospitals, transfusion centres, etc.) and free of charge. Nevertheless, vast blood resources exist worldwide and their exploitation will continue for many years to come. However, blood remains the ideal option for retrieving an immense amount of genetic and biochemical information from a few individuals.

2.4.2. Solid Tissues

Solid tissues, such as tumour biopsies or skin, constitute another common type of resource with high DNA content readily available to some hospital-linked laboratories. However, solid tissues present more severe drawbacks than blood in terms of personnel, equipment and labour, and are immediately excluded from the panel of general de novo sampling strategies. Tumours usually represent uncontrolled cell proliferation, which may be the cause or the effect of somatic mutations. They may therefore be ideal samples for somatic and cancer genetics but they may unnecessarily complicate interpretation of genetic variation at the population level.

2.4.3 Cell Lines

The same comment applies to immortal, or immortalised, cell lines. Human cells are immortalised by transfection with viruses (e.g. Epstein-Barr) or genetically engineered agents. Despite their cost in labour and storage (usually in liquid nitrogen), they are considered an unlimited source of stable (non-degrading) DNA and of viable cells on which to perform biochemical and molecular studies long after the donors have died (Austin et al. 1996). However, immortal cell lines proliferate in vitro far faster than tumours do in vivo. Mutations and major chromosomal rearrangements accumulate at every generation. The classical commercial cell lines (e.g. HeLa, MCSF, etc.) have been grown for hundreds or thousands of generations. Their unnatural ploidies are only statistical estimates because of the enormous heterogeneity of karyotypes observed in a flask. A clear distinction should, therefore, be made between diploid cell isolates with a limited life expectancy until senescence and immortal lines capable of indefinite proliferation. Whereas the former may adequately represent the genome of the donor, the latter may carry unacceptable amounts of artificial genetic variation and may not be recommended for use in population studies.

2.4.4 Rare Specimens

The successes of PCRs have encouraged researchers to develop DNA extraction and amplification methods for traces of tissues and tissue remains that contain few cells or no cells at all. Among the most impressive achievements are the extraction, amplification and genetic analysis of DNA from fossils (Sidow et al. 1991; Miskin et al. 1998; Poinar et al. 1998) and from old, archaeological and prehistoric human remains (Paabo 1989; Krings et al. 1997, 1999). DNA has also been successfully amplified from paraffin-embedded histological collections, and from tissues fixed in formalin, ethanol, or other fixatives (Jackson et al. 1990; Barrios et al. 1992; Fiallo et al. 1992; Soler et al. 1992; Dillon et al. 1996; Pavelic et al. 1996; Longy et al. 1997; Koch et al. 1998). Forensic science has made equally impressive progress with traces of human cells from saliva, saliva stains on cigarettes or stamps, bloodstains, hairs with visible follicles or hair shafts, dandruff, semen or urine (Schreiber et al. 1988; Walsh et al. 1992; Fridez and Coquoz 1996; van Schie and Wilson 1997; Lee et al. 1998; Schneider 1998; Lorente et al. 1998; Yokota et al. 1998). Non-human problematic specimens include single feathers, fish scales and museum collections (Thomas et al. 1990; Cann et al. 1993; Dick et al. 1993; Thomas and Paabo 1993).

These techniques are irrelevant to population genetics and will not be covered here for two reasons. Firstly, they all require special preparative procedures, and, secondly, individuals that consent to participate in a genetic study are willing to donate more than one hair but will never donate a bone for DNA extraction. However, the above reports do support the argument that basic genetic analysis can be performed with very minimal samples. Therefore, a small amount of tissue is probably all that future geneticists will require to carry out multiple complex studies.

2.4.5 Mouth Mucosa

Mouth mucosa may become the principal DNA source for the large-scale genetic epidemiology of the future. Sterile cotton swabs or brushes are used to gently scrape interior mouth surfaces (Thomson et al. 1992; Ilveskoski et al. 1998). Alternatively, the subject swirls 5–10 ml of water or saline and spits the mouthwash into a plastic universal vial (Lench et al. 1988; Lum and Le Marchand 1998). Concentrated saliva has also been used (van Schie and Wilson 1997). Either way, the samples are stable for at least 1 week at room temperature and can be returned to the laboratory by post. Mouth mucosa constitutes a minimal biohazard and requires minimal care in handling.

Among the three techniques of sampling mouth mucosa discussed above, I favour mouthwashes. This requires no special sterile accessories. It provides about 10 times as many cells as either swabs or brushes. Removing cells from swabs or brushes represents an additional preparative step whereas a mouthwash is already a cell suspension.

2.5 A Fundamental Change in Sampling Methodology

What effort does it really take to suspend cells from swabs or brushes into a centrifuge vial? For the historical reasons presented above, molecular biologists are trained to respect a code of practice that is proper to chemistry and biomedical science. Everything on the molecular biology bench, including disposable plasticware, needs to be sterile. All water used in reactions needs to be of the highest purity and autoclaved. There is always a bucket of ice for the concern or thermal instability of reagents ("better keep it cool than discover later that it has gone off"). The invisibility of the chemicals in solution adds mystery. It takes much more time to question and adapt a technical protocol than to apply it religiously ("anyway, if it doesn't work it won't be my fault"). In the context of the general effort in performing sound molecular biology, transferring cells from a solid surface into suspension does not require special effort; it is rather a reflex. However, suspending cells from 20,000 brushes or swabs starts to become considerable.

An original protocol for buccal tissue sampling (Thomson et al. 1992; Steinberg et al. 1997) required that buccal cells be collected with sterile cotton swabs, smeared onto alcohol-cleaned (microscope) slides and allowed to dry. The slides would then be transported in special contamination-free holders to prevent breakage and removed from the slide using a (second) sterile cotton swab wet in sterile saline. Cells would be concentrated by centrifugation, suspended in buffer and stored at –20 °C until lysed. In our case, the cost of this operation would have to be multiplied by 20,000. What if the donors received an equivalent volume of pharmacy-standard cotton wool with which to wipe their cheeks and a plastic bag to keep the cells moist on their way to the laboratory, where the cells would be immediately lysed? I do not know. But for saving on: 40,000 swabs; 20,000 glass slides (and the associated risk of accident); 20,000 special holders; chemicals and storing vials; an autoclave; a centrifuge; at least one dedicated freezer; and on invaluable losses of cells at each transfer step – all before DNA extraction – it seems worth trying.

New methods developed with throughput and cost in mind are urgently required. Every step of existing protocols needs to be titrated down to elimination. A clear distinction should be made between convenience and necessity. Is a 30-min incubation necessary for the completion of a reaction or only for a coffee break? The precise role of each chemical needs to be (re-)investigated against cheaper or safer alternatives. Should a mouthwash be in (sterile) saline, distilled water or tap water? By reflex, based on the common experience with blood cells, and with cell-culture media that are all isotonic (containing 0.9% NaCl), saline would be the choice. It is true that human cells would sooner or later burst from osmotic forces in a hypotonic environment or would be dehydrated in a hypertonic medium. In a pilot study (Spanakis, unpubl.) involving about 100 students, half of the subjects tried saline and half water. Without exception, saline was voted against as unpleasant or extremely unpleasant, a sign that mouth tissues were unhappy in the salty environment. Only then, did water seem an obvious solution: this is what mouth tissues experience several times per day. Molecular biologists always use sterile, distilled water for the perfect control and repeatability of chemical reactions. Nevertheless, neither saliva (the natural environment of buccal tissues) nor drinking water is sterile or distilled. In addition, a subject would more comfortably take the water he/she is familiar with than a preparation a scientist offers. The psychological effect may be minute, yet avoidable. Indeed, cells collected for the pilot experiment withstood water better than saline. The same reasoning should be carried through every step of DNA extraction to banking.

2.6 DNA Isolation and Purification

Of course, the precise method of DNA extraction largely depends on the start-ing tissue. A Medline search under "DNA/isolation and purification" would now list more than 22,000 entries, since 1965, with an average of ~1,500 entries per year in the past 10 years. These numbers reflect the breadth of biological sources and extraction variations used. I cannot pretend to be familiar with all of them, neither will I try to narrow down the subject by excluding, for example, methods devised for non-human sources or for sources that I con-sider unsuitable for large-scale epidemiology (solid tissues, rare, awkward, archival or forensic specimens). Although individual protocols are very rapidly being superseded, as the above numbers suggest, their constituent steps may remain valid and relevant. These can be re-combined towards ever faster and more cost-effective ways to collect and store DNA. Indeed, for such purposes, there may be no difference between a mouthwash, an immortal cell line, a pro-tozoan culture or a suspension of small soft invertebrates (Dick et al. 1993).

DNA isolation usually proceeds in five steps: preparation of the cells, lysis, removal of proteins and lipids, precipitation of DNA and re-suspension.

2.6.1 Preparation

The procedures of the first step vary enormously with the source tissue. To detach cells of a solid tissue from each other and make them accessible to lysing agents the tissue is usually ground, chopped, pulverised, sonicated, vortexed or treated with trypsin or other proteinases that digest cell adhesion proteins. For mechanical destruction, a soft tissue may need to be hardened first by drying or freezing in liquid nitrogen. Cultured adherent cells are usually detached from the plastic container with trypsin or scraping but detachment may not always be necessary. Mechanical forces may break cells and release endogenous DNases that could reduce the yield and integrity of DNA. Any proteinases used, on the other hand, must be diluted by washing the cells in a special isotonic medium (phosphate-buffered saline, PBS) and concentrating them again by centrifugation. Otherwise, the lysates contain too much added protein that can complicate DNA extraction.

Protocols for blood usually include extensive preparation to isolate and con-centrate nucleated (white) cells and remove the erythrocytes. Cell separation is based on the differential densities of white and red cells. A dense sugar solu-tion (usually ficoll or a commercial preparation) is carefully added beneath the blood so that the two phases do not mix. After gentle centrifugation (typically for about 10 min) the white cells sit on the interface between the dense bottom phase and the serum whereas the red cells sink through the dense phase to the bottom of the tube. If the interface is too red after the first centrifugation, a second round of cell separation may be required. The interface is re-suspended this time in an adequate volume (usually the same as that of the blood processed) of PBS. The white cells are eventually removed with the aid of yet another pipette, transferred in PBS and centrifuged once more to dilute out traces of serum and/or separation medium. Blood laboratories routinely purchase large stocks of special centrifugation tubes and disposable plasticware (pipettes) that commonly occupy a dedicated storeroom.

As blood is by far the commonest single source of human DNA, the variations in tissue preparation are so many that they would alone fill the whole of this chapter (Visvikis et al. 1998). The main aims are to adapt the procedures to the various ways of preserving blood and to reduce the length and/or cost of this extremely tedious procedure. There are, for example, adaptations for blood stored dried on Guthrie cards (Schneeberger et al. 1992; Makowski et al. 1998); for whole blood (Gustincich et al. 1991; Parzer and Mannhalter 1991; Lahiri et al. 1992; Weisberg et al. 1993); for blood containing different anti-coagulants such as EDTA, citrate or heparin (Gustafson et al. 1987); for blood preserved with urea (Gelhaus et al. 1995); as well as for clotted blood (Everson et al. 1993) or serum alone (Dixon et al. 1998).

Rare specimens, archival material and awkward tissues obviously require special preparation before DNA extraction. Mouthwashes require no preparation.

2.6.2 Cell Lysis

The cells are usually lysed with the aid of detergents like sodium dodecyl sulphate (SDS), Nonidet P40 or Triton X-100. Some protocols include separate steps for rupturing cell and nuclear membranes but, in most cases, DNA is instantly released upon suspension of cells in a lysis buffer. This is probably why the lysis step has not evolved much over the years. However, developers of new lysis strategies must take into account the possible interference of each ingredient with subsequent steps of extraction and analysis and with the stability of the DNA in the long term. For example, traces of SDS or of other detergents may severely inhibit the polymerase used in the PCR. Some salts may, at certain concentrations, (partially) co-precipitate with DNA and affect its stability during storage.

Mechanical methods such as vortexing or sonication aid lysis and may significantly increase the yield. For fear of breaking the DNA into pieces, however, the original protocols strongly discouraged the use of such methods. Every mixing step in the extraction procedure was, therefore, carried out by gentle repeated inversions of the extraction tube, and mixing alone could account for several hours of labour if special machines had not flooded the market. It would evolve that even the most severe mechanical disruption of DNA results in fragments sufficiently long for most analyses, and certainly so for genotyping. Mechanical methods may be considered an alternative to biochemical lysis.

2.6.3 Purification

The question of purification relates to what endogenous cellular, or added, ingredients need to be removed from the lysate. The answer depends on what analytical techniques will be used and when. Researchers frequently go out of their way to produce highly purified high molecular weight DNA that they use in a single PCR and, then, throw away. At the other extreme, the risk is to over-simplify extraction and purification and find the extracts unusable after a few analyses.

Traditionally, attention has focused on the total protein of the lysate. The total protein content is estimated from the absorption by the extract of UV light of 280 nm wavelength. The extract is considered pure if the ratio of absorption at 260 nm (peak of the absorption spectrum of nucleic acids) over that at 280 nm is between 1.6 and 2.0, ideally 1.7. Values higher than 2.0 indicate degradation of DNA or chemical contamination whereas values below 1.6 indicate the presence of protein. However, only some proteins must be eliminated. Most proteins are innocuous for many types of analysis. DNase catabolises DNA. Histones are bound to eukaryotic chromosomes and may restrict access of other proteins. Proteinases catabolise proteins and are undesired in analytical procedures involving enzymes, such as restriction analysis or PCR. The early protocols invariably began purification with a protein digestion step. This was an over-night incubation of the lysate with added proteinase K. The primary purpose was to destroy DNase (and RNase for RNA extraction) at the start, and not to remove proteins or merely lyse nuclei as many people might think. Proteins, including proteinase K, were subsequently removed in several organic extraction steps with phenol and chloroform (Smith et al. 1970), sometimes mixed with isoamyl alcohol. An alkaline pH, usually around 8, ensures that the RNA is removed during selective DNA extraction. However, some protocols include an incubation of the lysate with RNase prior to extraction (Johns and Paulus-Thomas 1989; van Schie and Wilson 1997) whereas other methods have been especially devised for parallel extraction of both nucleic acids (Macnab et al. 1988; Jackson et al. 1990; Raha et al. 1990).

Unlike RNase, DNase is extremely sensitive to high temperatures and a few minutes incubation of the lysate at 55 °C, or higher, completely and irreversibly destroys this enzyme. Furthermore, there is no point in destroying enzymes that are to be eliminated anyway. A wealth of simpler methods has begun, therefore, to accumulate since the late 1980s. These contained no enzymatic digestion or organic extraction. One family of methods, known as salting-out methods, employs high concentrations of salts such as sodium chloride (Gilden et al. 1982; Mullenbach et al. 1989; Lahiri et al. 1992), ammonium chloride (Topic and Gluhak 1991), sodium perchlorate (Wilcockson 1973, 1975; Johns and Paulus-Thomas 1989), lithium chloride (Raha et al. 1990) or potassium acetate (Potter et al. 1985; Sokolov et al. 1989) to precipitate the proteins. Another family, known as solid-phase extraction methods, uses chromatography to separate DNA from everything else in crude lysates (McCormick 1989; Scherczinger et al. 1997; Yang et al. 1998).

These methods use the differential affinities of DNA and impurities for the resin, affinity gel, or silica in the column but most solid-phase applications have, so far, been analytical rather than preparative (Klug and Famulok 1994; Labrou and Clonis 1994; Novotny 1997). At first, chromatography was too laborious and too expensive to be considered for high-throughput applications. Now, commercially available kits designed for simultaneous purification of 96 lysates directly into industry-standard 96-well arrays (e.g. QIAGEN, http://www.qiagen.com; see also http://news.bioresearchonline.com for other commercial DNA extraction solutions) have substantially higher throughput but remain too expensive in consumables for large-scale projects. DNA is finally precipitated with sodium acetate (if the lysate does not already contain added salts) and ethanol. The pellet may be washed once with 75% ethanol and re-suspended in buffer or water before use.

It would appear from this section that protein digestion, RNA digestion, protein precipitation, organic extraction, and chromatography are not indispensable. Kejnovsky and Kypr (1997, 1998) reported a method of DNA sedimentation using millimolar concentrations of zinc chloride, a relatively inexpensive salt, combined with sub-millimolar concentrations of phosphate. Perhaps a selective precipitation of DNA is the only essential step for genetic analysis; or is DNA extraction essential at all? Gross and Rotzer (1998) have reported that they were able to perform PCR from 20 µl of whole blood processed in a single 1.5-ml reaction tube without digesting proteins, precipitating DNA or even lysing the nuclei.

In recent years, DNA isolation has been reduced from an over-night procedure to one requiring just a few minutes. The signs are that this evolution is accelerating. An interesting comparison of the yield against the speed of various methods was made some years ago (Lahiri et al. 1992). These authors claimed of course that their method (salting out) was not only the most rapid but also the most productive. Their results, looked at more closely, can be interpreted in an even more optimistic way: there was a positive statistical trend between speed and yield; the faster the method (the fewer and shorter the steps of extraction), the higher the yield. It is likely that DNA is lost at every step of extraction and for as long as the extract remains impure, degradation continues.

Tradition is difficult to change. Even those most concerned (in this case molecular genetic epidemiologists) remain either unaware of, or reluctant to adopt new methods, particularly while projects are running. Many, if not most, laboratories claiming to perform large-scale genetic epidemiology are still using DNA sampling methods devised more than 10 years ago. However, the most important factor that delays acceptance of new methods of tissue sampling and DNA extraction is, in my opinion, the almost complete lack in the literature of evaluation data on the stability of samples in the long term. This problem is more acute with new methods that have not had the time to be tested in the field.

2.7 Sample Storing

The first relevant question is whether one should store the samples or immediately extract and store the DNA instead. The answer depends on the circumstances and the needs of each study. From some of our ongoing studies we have acquired frozen blood, for some frozen DNA, and we have no option. For our new projects, where we will be looking in depth into the genetics of a few samples, we will have to take a decision. However, for our 20,000-women study mentioned above, which more closely represents the routine genetic epidemiology of the future, we have again no option. As the study requires no biochemical measurements we have decided to collect one mouthwash per subject in a 25-ml universal vial. Since mouthwashes are not sterile, they cannot remain at room temperature for the duration of the study and beyond. To freeze 20,000 universal vials we would need 7 large freezers, 1–2 dedicated air-conditioned rooms, probably fitted with over-heating alarms, 20,000 freezer-proof labels and a fairly sophisticated rack/box system for easy access and safe handling. We had limited funds for all of these and we decided to spend it on research rather than on storage. Therefore, we could store DNA or nothing.

2.7.1 Storing Tissues

There are several studies of long-term stability of human tissues in the context of genetic testing; most of them on blood (Gustafson et al. 1987; Madisen et al. 1987; Polakova et al. 1989; Ross et al. 1990; Arseniev et al. 1992; Lahiri and Schnabel 1993; Weisberg et al. 1993; Muralidharan and Wemmer 1994; Gelhaus et al. 1995; Austin et al. 1996; Visvikis et al. 1998). There is general agreement that, in whatever form, blood, like mouthwashes or other common tissues, can withstand any temperature (e.g. room temperature) for a few days or weeks (Aggarwal et al. 1992; Muralidharan and Wemmer 1994); however, in tropical climates, immediate preservation is recommended (Bates et al. 1991). Freezing or drying is also generally accepted for long-term preservation. Blood stored for more than 20 years yields "sufficient" amounts of "good" DNA (Madisen et al. 1987; Ahmad et al. 1995).

The debated question is whether and to what extent frozen blood, or cells in general, can withstand repeated thawing–freezing cycles without losing in amount or quality of the DNA eventually extracted (Gustafson et al. 1987; Ross et al. 1990; Nguyen et al. 1991; Arseniev et al. 1992; Lahiri and Schnabel 1993; Ahmad et al. 1995; Visvikis et al. 1998). The results are often contradictory and inconclusive; compare, for example, Gustafson et al. (1987) with Ross et al. (1990). There is, however, no systematic study with formal statistical analysis. Between-laboratory comparisons are yet impossible because of the wide variety of analytical methods, standards, criteria and definitions of "quantity" and "quality" used.

In theory, water crystals growing during freezing may mechanically break some cells and release DNase and other hydrolytic factors. As soon as the sample returns to, and for as long as it remains at, room temperature released DNase will damage DNA. The extent of such damage will depend not only on the number of freezing cycles (number of damaged cells) but also on the total time for which samples have been left thawed before DNA extraction. Detection of degradation by electrophoresis or by spectrophotometry does not mean that all DNA has been degraded and, indeed, the sample may still be perfect for most analyses. The cost of degradation should finally be compared with the cost of aliquoting. The same principle of relativity applies to the costs of storing frozen cell suspensions against the costs of the special procedures and effort required for extracting equivalent amounts of DNA from dried samples.

2.7.2 Storing DNA

DNA is a reasonably stable substance. The fact that DNA has been isolated from fossils means that, given the right conditions, it can remain intact for longer than a lifetime. In the laboratory, DNA solutions have withstood room temperature for at least many months and, under refrigeration or dry, for several years (Madisen et al. 1987; Ross et al. 1990). This does not mean, of course, that all DNA preparations are equally stable. Traces of microbial contamination producing DNase are capable of digesting DNA very rapidly. The lower the temperature of storage, the slower such activity should be. There are fewer studies of DNA stability in long-term storage than there are for tissues. However, the few studies I have come across are a lot more convincing. One of them tested frozen or dried DNAs prepared more than 40 years ago and found them intact, although, and this is a common experience, dry DNA is difficult to re-dissolve (Madisen et al. 1987).

Most of what we know about the degradation of DNA in solution or in a dry condition comes from carefully controlled short-term chemical studies rather than long-term observations. Enzymatic or chemical postmortem hydrolysis is considered an important cause of degradation but much of this is supposed to be stopped by the purification procedure. Another major cause is oxidation. Atmospheric oxygen and probably endogenous oxygen donors, such as metal oxides or protein-associated superoxides, oxidise DNA fairy rapidly (oxidation detectable within months) leading to degradation. The oxidation, however, is not direct but is mediated by lipids. DNA from dry tissues (containing lipids) stored in nitrogen or argon atmospheres continued to degrade though at much slower rate than under oxygen. Prior extraction of lipids rendered DNA resistant to oxidation and degradation even in oxygen. Bacteriophage lambda DNA (naturally lipid-free) was also stable under all conditions (Matsuo et al. 1995). I have never encountered an extraction quality test that measured lipids in DNA extracts to be stored long term. Maybe, protocols that avoid organic extraction seriously jeopardise the long-term stability of DNA and, most probably, so does storing tissues rather than extracts.

Another set of relevant chemical studies concerns the effects of ethanol, commonly used to precipitate nucleic acids. We know that unless stored in ethanol RNA rapidly degrades. Because DNA is fairly stable in solution (water or buffer) in the short term, its relative stability as a precipitate in ethanol has not been of particular interest so far. However, recent physicochemical and structural studies have revealed dramatic effects of ethanol on the DNA structure, which may prove relevant to long-term storage. Ethanol, alone or with salts (including sodium chloride or perchlorate), causes DNA to adopt A-, P-, X-, or Z-conformations that are more rigid and compact than its usual B-conformation observed in aqueous solution. The exact change and the stability of the adopted structure depend on the ionic strength and/or on the presence of metal ions such as Mg, Ca, Co, Cs, Fe, Ni, and Zn (Vorlickova et al. 1991a,b; Arscott et al. 1995; Piskur and Rupprecht 1995; Cheatham et al. 1997; Pospisilova and Kypr 1998). A massive contraction, dehydration, an increase in hydrophobicity and, eventually, precipitation accompany these conformational changes (Schultz et al. 1994). In this way, ethanol may protect the molecule from oxidation, enzyme attacks and other degradation processes (Saez et al. 1993; Hiramoto et al. 1995; Jacobs and Neilan 1995; Dillon et al. 1996; Frisman et al. 1997; Nakao and Augusto 1998). Certainly, more studies focused on the stability of DNA in ethanol are needed.

2.8 Sample Banking

The term sample banking means a lot more than a few stored specimens, whether tissues or DNA extracts. It means a purposely organised collection where specimens and their accompanying data can be added and retrieved by others. Provision must be made for safety, security, clarity, confidentiality and error proofing for both specimens and data. The system is subject to an explicit quality control and to good record keeping.

For example, technical protocols of sampling through to analysis and distribution of the collection as well, as all correspondence with the funding and ethics organisations and all relevant publications, must be readily available to potential users. On the other hand, access to the identity of the individuals should be strictly restricted. Some banks may, however, be anonymous or made anonymous. Samples must be kept in duplicate at different locations and securely locked. Records and accompanying data must be regularly updated and backed-up as several physically independent electronic copies at different locations as well as in hard copies for easy consultation. Appliances such as freezers or other equipment and accessories associated with the collection (e.g. plasticware, racks, boxes, etc.) must be safety certified, regularly serviced and of sound quality for the safety of both the samples and the personnel. DNA samples stored in ethanol should be treated as highly flammable and be kept in inflammable containers.

A restricted committee manages the bank of samples we are creating from a local population. The samples are kept in 96-well arrays and are identified by the array number and their coordinates. Labelling materials, labour and the probability of error are thus reduced 100-fold. For access, potential users must submit a written proposal providing a sort description of their project and evidence of financial support and ethical acceptance. Validation of the proposed methods of using a pilot set of samples prepared in an identical way (by our laboratory) is also required. Proposals are peer-reviewed.

The accompanying database has built-in error tracking mechanisms and is accessible only to authorised "visitors" and "users" with individual passwords. Visitors can only consult the database, whereas users can also update it. The program records consultations and updates with the user-name, date and time of each event. Users have "signatures" and can only update the database by "signing". Signed entries are automatically locked and the user can only correct any subsequently traced errors in the presence of the manager. A "receipt" of the entries is printed for the user upon expressing intention to update the database. The user can thus double-check the entries before signing. Signing the program automatically updates the master database and produces an electronic backup on the user's floppy disk as well as a hard copy of the latest array of data. For clarity, each field of data is printed in an 8×12 array format that serves as a map for finding the samples. The users' floppy disks are automatically updated, and locked when full. The system must be user-friendly for computer-illiterate personnel and has proven efficient. This automation also makes the database resistant to human error and inaccessible to deliberate corruption.

To operate conservatively with a bank that contains a finite resource of DNA, there are several techniques to consider. When, for example, a user intends to study numerous loci dispersed throughout the genome, the product of a whole-genome pre-amplification – whereby a large random proportion of the genome is amplified by PCR using degenerate primers, e.g. Zhang et al. (1992) – can be used instead of the original genomic DNA. Long-segment pre-amplification (Ohler and Rose 1992; Cheng et al. 1994) provides an equivalent secondary template for further, nested amplification of closely linked loci. For loci that cannot be pre-amplified, more sensitive analytical methods should be considered (e.g. radioactive labelling) to reduce the amount of original template to be consumed.

The history of blood banks is almost as long as that of transfusion. Reports of other human tissue banks began to accumulate in the 1970s. However, the vast majority of human biological specimen collections are built for purposes other than genetic analysis. Until recently, the term "DNA bank" has almost exclusively been used as a synonym of "DNA library" or "database". The first human DNA bank for population genetics I could trace was reported in 1983 and comprised 50 specimens (Rahuel et al. 1983). Today, the intersection of

MeSH terms: "DNA/isolation and purification", "/analysis", "/diagnostic use" or "/genetics" *and* "Biological Specimen Banks" – i.e. rigorously organised collections – contains only a handful of relevant reports. Banks created or used for the study of human genes that I have not already mentioned can be found in Greenberg et al. (1993), Kuykendall and Lory (1995), Akaza (1997), Pero et al. (1998) and Chen and Zhang (1998). This is not to say that such banks are rare. Practically every diagnostic centre and every genetic research laboratory have their own collections. These "private banks" are, however, unavailable to others, either because they were created without provisions for sharing or simply because they remain unpublished. A survey of academic and commercial laboratory DNA banks by McEwen and Reilly (1995) revealed that although DNA banking activity is growing, most academically based (potentially available to others) banks lack appropriate documentation.

Under "DNA" and "banking" text terms the reader will find, nevertheless, a wealth of philosophical essays, discussions, reviews, perspectives and guidelines, mostly on ethical, but also on methodological issues (Anonymous 1988, 1989a,b; Yates et al. 1989; Hall et al. 1991; Cann et al. 1993; Knoppers 1993, 1999; Lyttle 1997; Henriksen and Horder 1998; Khoury and Yang 1998; Knoppers et al. 1998). The phenomenon of having more review-type articles than primary research reports to read is typical of a discipline at birth. DNA banks will, no doubt, be the biological materials of future population genetics. We have only just begun to think how we will create them. But, unlike science, ethical rules are not universal. The discussion should never reach conclusion. Scientists can only anticipate that ethics, legislation and culture in general will keep balancing profit and harm and will keep evolving at the right speed so as to accommodate, and lead, genetic advances.

References

Anonymous (1968) Nobel Prizes for Medicine, 1968. Nature 220:324–325

Anonymous (1988) DNA banking and DNA analysis: points to consider. Ad Hoc Committee on DNA Technology, American Society of Human Genetics. Am J Hum Genet 42:781–783

Anonymous (1989a) DNA banking and DNA analysis: points to consider. J Health Hosp Law 22:282–283

Anonymous (1989b) Gene testing in autosomal dominant polycystic kidney disease: results of National Kidney Foundation workshop. Scientific Advisory Board of the National Kidney Foundation. Am J Kidney Dis 13:85–87

Aggarwal RK, Lang JW, Singh L (1992) Isolation of high-molecular-weight DNA from small samples of blood having nucleated erythrocytes, collected, transported, and stored at room temperature. Genet Anal Tech Appl 9:54–57

Ahmad NN, Cu-Unjieng AB, Donoso LA (1995) Modification of standard proteinase K/phenol method for DNA isolation to improve yield and purity from frozen blood. J Med Genet 32: 129–130

Akaza T (1997) HLA typing in the Japanese Bone Marrow Program (in Japanese). Rinsho Byori 45:157–160

Andrieu N, Goldstein AM (1998) Epidemiologic and genetic approaches in the study of gene-environment interaction: an overview of available methods. Epidemiol Rev 20:137–147

Angelico F, Del Ben M (1992) Reliability and power of genetic studies. Ann Ist Super Sanita 28: 409–413

Arscott PG, Ma C, Wenner JR, Bloomfield VA (1995) DNA condensation by cobalt hexaammine(III) in alcohol-water mixtures: dielectric constant and other solvent effects. Biopolymers 36:345–364

Arseniev L, Battmer K, Andres J, Poliwoda H, Stangel W, Link H (1992) Influence of the freezing-thawing-washing procedure on the cytomegalovirus DNA content of cellular blood products. Infusionsther Transfusionsmed 19:199–201

Austin MA, Ordovas JM, Eckfeldt JH, Tracy R, Boerwinkle E, Lalouel JM, Printz M (1996) Guidelines of the National Heart, Lung, and Blood Institute Working Group on blood drawing, processing, and storage for genetic studies (published erratum appears in Am J Epidemiol 1997, 145(6):570). Am J Epidemiol 144:437–441

Bacon BR, Schilsky ML (1999) New knowledge of genetic pathogenesis of hemochromatosis and Wilson's disease. Adv Intern Med 44:91–116

Baltimore D (1970) RNA-dependent DNA polymerase in virions of RNA tumour viruses. Nature 226:1209–1211

Barrios L, Miro R, Corominas M, Pellicer A, Egozcue J (1992) A method to extract DNA for molecular studies from cells fixed in Carnoy. Cancer Genet Cytogenet 59:217–218

Bates I, Bedu-Addo G, Rutherford TR (1991) Extracting, storing, and transporting whole blood DNA under tropical conditions. J Clin Pathol 44:605–606

Bradley DM, Parsons EP, Clarke AJ (1993) Experience with screening newborns for Duchenne muscular dystrophy in Wales (see comments). Br Med J 306:357–360

Brown PO, Botstein D (1999) Exploring the new world of the genome with DNA microarrays. Nat Genet 21:33–37

Cann RL, Feldman RA, Freed LA, Lum JK, Reeb CA (1993) Collection and storage of vertebrate samples. Methods Enzymol 224:38–51

Cavalli-Sforza L (1998) Man and the diversity of his genome. An extraordinary phase in the history of population genetics. Pathol Biol (Paris) 46:98–102

Cavalli-Sforza LL, Wilson AC, Cantor CR, Cook-Deegan RM, King MC (1991) Call for a worldwide survey of human genetic diversity: a vanishing opportunity for the Human Genome Project. Genomics 11:490–491

Cheatham TE III, Crowley MF, Fox T, Kollman PA (1997) A molecular level picture of the stabilization of A-DNA in mixed ethanol-water solutions. Proc Natl Acad Sci USA 94:9626–9630

Chen Z, Zhang S (1998) Chinese Human Genome Project – opportunity and challenge (in Chinese). Chung Hua I Hsueh I Chuan Hsueh Tsa Chih 15:195–197

Cheng S, Fockler C, Barnes WM, Higuchi R (1994) Effective amplification of long targets from cloned inserts and human genomic DNA. Proc Natl Acad Sci USA 91:5695–5699

Day INM, Spanakis E, Palamand D, Weavind GP, O'Dell SD (1998) Microplate-array diagonal-gel electrophoresis (MADGE) and melt-MADGE: tools for molecular-genetic epidemiology. Trends Biotechnol 16:287–290

Dick M, Bridge DM, Wheeler WC, DeSalle R (1993) Collection and storage of invertebrate samples. Methods Enzymol 224:51–65

Dillon N, Austin AD, Bartowsky E (1996) Comparison of preservation techniques for DNA extraction from hymenopterous insects. Insect Mol Biol 5:21–24

Dixon SC, Horti J, Guo Y, Reed E, Figg WD (1998) Methods for extracting and amplifying genomic DNA isolated from frozen serum. Nat Biotechnol 16:91–94

Elston RC (1998a) Methods of linkage analysis and the assumptions underlying them (see comment). Am J Hum Genet 63:931–934

Elston RC (1998b) Linkage and association. Genet Epidemiol 15:565–576

Everson RB, Mass MJ, Gallagher JE, Musser C, Dalzell J (1993) Extraction of DNA from cryopreserved clotted human blood. Biotechniques 15:18–20

Fiallo P, Williams DL, Chan GP, Gillis TP (1992) Effects of fixation on polymerase chain reaction detection of *Mycobacterium leprae*. J Clin Microbiol 30:3095–3098

Flanders WD, Khoury MJ (1996) Analysis of case-parental control studies: method for the study of associations between disease and genetic markers (published erratum appears in Am J Epidemiol 1997, 145(5):477–478). Am J Epidemiol 144:696–703

Fridez F, Coquoz R (1996) PCR DNA typing of stamps: evaluation of the DNA extraction (see comments). Forensic Sci Int 78:103–110

Frisman EV, Zarubina OP, Zyrianova IM, Kukhtin AA, Tarasenko OA (1997) The role of ethanol during gamma-irradiation of water-salt DNA solutions. Biofizika 42:117–124

Gelhaus A, Urban B, Pirmez C (1995) DNA extraction from urea-preserved blood or blood clots for use in PCR (published erratum appears in Trends Genet 1995, 11(4):129). Trends Genet 11:41

Gellert M, Little JW, Oshinsky CK, Zimmerman SB (1968) Joining of DNA strands by DNA ligase of *E. coli*. Cold Spring Harbor Symp Quant Biol 33:21–26

Gilden DH, Shtram Y, Friedmann A, Wellish M, Devlin M, Cohen A, Fraser N, Becker Y (1982) Extraction of cell-associated varicella-zoster virus DNA with Triton X-100-NaCl. J Virol Methods 4:263–275

Grann VR, Whang W, Jacobson JS, Heitjan DF, Antman KH, Neugut AI (1999) Benefits and costs of screening Ashkenazi Jewish women for BRCA1 and BRCA2. J Clin Oncol 17:494–500

Greenberg J, Bartmann L, Ramesar R, Beighton P (1993) Retinitis pigmentosa in southern Africa. Clin Genet 44:232–235

Gross M, Rotzer E (1998) Rapid DNA extraction method for genetic screening. Eur J Med Res 3:173–175

Gussin GN, Capecchi MR, Adams JM, Argetsinger JE, Tooze J, Weber K, Watson JD (1966) Protein synthesis directed by DNA phage messengers. Cold Spring Harbor Symp Quant Biol 31: 157–171

Gustafson S, Proper JA, Bowie EJ, Sommer SS (1987) Parameters affecting the yield of DNA from human blood. Anal Biochem 165:294–299

Gustincich S, Manfioletti G, Del Sal G, Schneider C, Carninci P (1991) A fast method for high-quality genomic DNA extraction from whole human blood. Biotechniques 11:298–300, 302

Hall J, Hamerton J, Hoar D, Korneluk R, Ray P, Rosenblatt D, Wood S (1991) Policy statement concerning DNA banking and molecular genetic diagnosis. Canadian College of Medical Geneticists. Clin Invest Med 14:363–365

Hall NE, Axelrod DE (1977) Interference of cellular ferric ions with DNA extraction and the application to methods of DNA determination. Anal Biochem 79:425–430

Harbers E (1969) On the discovery of DNA by Friedrich Miescher 100 years ago. Ger Med Mon 14:517–518

Henriksen FL, Horder M (1998) DNA biobanks. Establishment and maintenance. Ugeskr Laeger 160:5025–5029

Hiramoto K, Kaku M, Kato T, Kikugawa K (1995) DNA strand breaking by the carbon-centered radical generated from 4-(hydroxymethyl) benzenediazonium salt, a carcinogen in mushroom *Agaricus bisporus*. Chem Biol Interact 94:21–36

Hood LE, Hunkapiller MW, Smith LM (1987) Automated DNA sequencing and analysis of the human genome. Genomics 1:201–212

Hotchkiss RD (1979) The identification of nucleic acids as genetic determinants. Ann NY Acad Sci 325:320–342

Hotchkiss RD (1995) DNA in the decade before the double helix. Ann NY Acad Sci 758:55–73

Houwen RH, Baharloo S, Blankenship K, Raeymaekers P, Juyn J, Sandkuijl LA, Freimer NB (1994) Genome screening by searching for shared segments: mapping a gene for benign recurrent intrahepatic cholestasis. Nat Genet 8:380–386

Hunkapiller MW, Hewick RM, Dreyer WJ, Hood LE (1983) High-sensitivity sequencing with a gas-phase sequenator. Methods Enzymol 91:399–413

Ilveskoski E, Lehtimaki T, Erkinjuntti T, Koivula T, Karhunen PJ (1998) Rapid apolipoprotein E genotyping from mailed buccal swabs. J Neurosci Methods 79:5–8

Jackson DP, Lewis FA, Taylor GR, Boylston AW, Quirke P (1990) Tissue extraction of DNA and RNA and analysis by the polymerase chain reaction. J Clin Pathol 43:499–504

Jacobs D, Neilan BA (1995) Long-term preservation of DNA in agarose gels using 70% ethanol. Biotechniques 19:892–894

Johns MB Jr, Paulus-Thomas JE (1989) Purification of human genomic DNA from whole blood using sodium perchlorate in place of phenol. Anal Biochem 180:276–278

Kejnovsky E, Kypr J (1997) DNA extraction by zinc. Nucleic Acids Res 25:1870–1871

Kejnovsky E, Kypr J (1998) Millimolar concentrations of zinc and other metal cations cause sedimentation of DNA. Nucleic Acids Res 26:5295–5299

Khorana HG (1965) Polynucleotide synthesis and the genetic code. Fed Proc 24:1473–1487

Khoury MJ, Beaty TH (1994) Applications of the case-control method in genetic epidemiology. Epidemiol Rev 16:134–150

Khoury MJ, Yang Q (1998) The future of genetic studies of complex human diseases: an epidemiologic perspective. Epidemiology 9:350–354

Kidd JR, Kidd KK, Weiss KM (1993) Human genome diversity initiative. Hum Biol 65:1–6

Klug SJ, Famulok M (1994) All you wanted to know about SELEX. Mol Biol Rep 20:97–107

Knoppers BM (1993) Confidentiality in genetic testing: legal and ethical issues in an international context. Med Law 12:573–582

Knoppers BM (1999) Status, sale and patenting of human genetic material: an international survey (in process citation). Nat Genet 22:23–26

Knoppers BM, Hirtle M, Lormeau S, Laberge CM, Laflamme M (1998) Control of DNA samples and information. Genomics 50:385–401

Koch DA, Duncan GA, Parsons TJ, Pruess KP, Powers TO (1998) Effects of preservation methods, parasites, and gut contents of black flies (Diptera: Simuliidae) on polymerase chain reaction products. J Med Entomol 35:314–318

Kornberg A, Lehman IR, Bessman MJ, Simms ES (1989) Enzymic synthesis of deoxyribonucleic acid. 1956 (classical article). Biochim Biophys Acta 1000:57–58

Krings M, Stone A, Schmitz RW, Krainitzki H, Stoneking M, Paabo S (1997) Neanderthal DNA sequences and the origin of modern humans (see comments). Cell 90:19–30

Krings M, Geisert H, Schmitz RW, Krainitzki H, Paabo S (1999) DNA sequence of the mitochondrial hypervariable region II from the Neanderthal type specimen (in process citation). Proc Natl Acad Sci USA 96:5581–5585

Kuykendall LH, Lory DS (1995) Teamwork and leadership: the development of the National Psoriasis Tissue Bank. Dermatol Nurs 7:298–301

Labrou N, Clonis YD (1994) The affinity technology in downstream processing. J Biotechnol 36:95–119

Lahiri DK, Schnabel B (1993) DNA isolation by a rapid method from human blood samples: effects of MgCl$_2$, EDTA, storage time, and temperature on DNA yield and quality. Biochem Genet 31:321–328

Lahiri DK, Bye S, Nurnberger JI Jr, Hodes ME, Crisp M (1992) A non-organic and non-enzymatic extraction method gives higher yields of genomic DNA from whole-blood samples than do nine other methods tested. J Biochem Biophys Methods 25:193–205

Lee HC, Ladd C, Scherczinger CA, Bourke MT (1998) Forensic applications of DNA typing, 2. Collection and preservation of DNA evidence. Am J Forensic Med Pathol 19:10–18

Lench N, Stanier P, Williamson R (1988) Simple non-invasive method to obtain DNA for gene analysis. Lancet 1:1356–1358

Linder MW, Valdes R Jr (1999) Fundamentals and applications of pharmacogenetics for the clinical laboratory (in process citation). Ann Clin Lab Sci 29:140–149

London SJ, Idle JR, Daly AK, Coetzee GA (1999) Genetic variation of CYP2A6, smoking, and risk of cancer (letter). Lancet 353:898–899

Longy M, Duboue B, Soubeyran P, Moynet D (1997) Method for the purification of tissue DNA suitable for PCR after fixation with Bouin's fluid. Uses and limitations in microsatellite typing. Diagn Mol Pathol 6:167–173

Lorente M, Entrala C, Lorente JA, Alvarez JC, Villanueva E, Budowle B (1998) Dandruff as a potential source of DNA in forensic casework. J Forensic Sci 43:901–902

Lucotte G, Mercier G (1998) Distribution of the CCR5 gene 32-bp deletion in Europe. J Acquir Immune Defic Syndr Hum Retrovirol 19:174–177

Lum A, Le Marchand L (1998) A simple mouthwash method for obtaining genomic DNA in molecular epidemiological studies. Cancer Epidemiol Biomarkers Prev 7:719–724

Lyttle J (1997) Is informed consent possible in the rapidly evolving world of DNA sampling? CMAJ 156:257–258

Macnab JC, Offord EA, Neilson L, Leake RE, Kitchener HC (1988) A technique for detecting both DNA and RNA in the same tissue biopsy sample. Nucleic Acids Res 16:11371

Madisen L, Hoar DI, Holroyd CD, Crisp M, Hodes ME (1987) DNA banking: the effects of storage of blood and isolated DNA on the integrity of DNA. Am J Med Genet 27:379–390

Makowski GS, Davis EL, Nadeau F, Hopfer SM (1998) Polymerase chain reaction amplification of Guthrie card deoxyribonucleic acid: extraction of nucleic acid from filter matrices. Ann Clin Lab Sci 28:254–259

Matsuo S, Toyokuni S, Osaka M, Hamazaki S, Sugiyama T (1995) Degradation of DNA in dried tissues by atmospheric oxygen. Biochem Biophys Res Commun 208:1021–1027

Maxam AM, Gilbert W (1977) A new method for sequencing DNA. Proc Natl Acad Sci USA 74: 560–564

McCormick RM (1989) A solid-phase extraction procedure for DNA purification. Anal Biochem 181:66–74

McEwen JE, Reilly PR (1995) A survey of DNA diagnostic laboratories regarding DNA banking. Am J Hum Genet 56:1477–1486

Melgar E, Goldthwait DA (1968a) Deoxyribonucleic acid nucleases. I. The use of a new method to observe the kinetics of deoxyribonucleic acid degradation by deoxyribonuclease I, deoxyribonuclease II, and Escherichia coli endonuclease I. J Biol Chem 243:4401–4408

Melgar E, Goldthwait DA (1968b) Deoxyribonucleic acid nucleases. II. The effects of metals on the mechanism of action of deoxyribonuclease I. J Biol Chem 243:4409–4416

Merikangas KR (1993) Genetic epidemiologic studies of affective disorders in childhood and adolescence. Eur Arch Psychiatry Clin Neurosci 243:121–130

Miskin I, Rhodes G, Lawlor K, Saunders JR, Pickup RW (1998) Bacteria in post-glacial freshwater sediments. Microbiology 144:2427–2439

Mullenbach R, Lagoda PJ, Welter C (1989) An efficient salt-chloroform extraction of DNA from blood and tissues. Trends Genet 5:391

Mullis K, Faloona F, Scharf S, Saiki R, Horn G, Erlich H (1986) Specific enzymatic amplification of DNA in vitro: the polymerase chain reaction. Cold Spring Harbor Symp Quant Biol 51: 263–273

Muralidharan K, Wemmer C (1994) Transporting and storing field-collected specimens for DNA without refrigeration for subsequent DNA extraction and analysis. Biotechniques 17:420–422

Nakao LS, Augusto O (1998) Nucleic acid alkylation by free radical metabolites of ethanol. Formation of 8-(1-hydroxyethyl)guanine and 8-(2-hydroxyethyl)guanine adducts. Chem Res Toxicol 11:888–894

Narang SA, Dheer SK, Michniewicz JJ (1968) A new general method for the synthesis of deoxyribopolynucleotides bearing a 5'-phosphomonoester and group. J Am Chem Soc 90:2702–2704

Newman B, Millikan RC, King MC (1997) Genetic epidemiology of breast and ovarian cancers. Epidemiol Rev 19:69–79

Nguyen C, Djabali M, Roux D, Jordan BR (1991) Very high molecular weight DNA for pulsed field gel studies can be obtained routinely from conventional frozen blood aliquots. Nucleic Acids Res 19:407

Nirenberg M (1965) Protein synthesis and the RNA code. Harvey Lect 59:155–185

Novotny MV (1997) Capillary biomolecular separations. J Chromatogr B Biomed Sci Appl 689: 55–70

Ohler LD, Rose EA (1992) Optimization of long-distance PCR using a transposon-based model system. PCR Methods Appl 2:51–59

Paabo S (1989) Ancient DNA: extraction, characterization, molecular cloning, and enzymatic amplification. Proc Natl Acad Sci USA 86:1939–1943

Parzer S, Mannhalter C (1991) A rapid method for the isolation of genomic DNA from citrated whole blood. Biochem J 273:229–231

Pavelic J, Gall-Troselj K, Bosnar MH, Kardum MM, Pavelic K (1996) PCR amplification of DNA from archival specimens. A methodological approach. Neoplasma 43:75–81

Pero RW, Olsson A, Bryngelsson C, von Schantz T, Simanaitis M, Sternby NH, Larsson SA, Elmstahl S, Janzon L, Berglund G (1998) Feasibility and quality of biological banking of human normal and tumor tissue specimens as sources of DNA for the Malmo Diet and Cancer Study. Cancer Epidemiol Biomarkers Prev 7:809–812

Piskur J, Rupprecht A (1995) Aggregated DNA in ethanol solution. FEBS Lett 375:174–178

Poinar HN, Hofreiter M, Spaulding WG, Martin PS, Stankiewicz BA, Bland H, Evershed RP, Possnert G, Paabo S (1998) Molecular coproscopy: dung and diet of the extinct ground sloth *Nothrotheriops shastensis* (see comments). Science 281:402–406

Polakova H, Kadasi L, Zelinkova M (1989) The yield and quality of DNA extracted from blood samples stored under various conditions. Bratisl Lek Listy 90:844–847

Pospisilova S, Kypr J (1998) UV light-induced duplex-to-duplex crosslinking of DNA molecules in aqueous ethanol solutions. Photochem Photobiol 67:386–390

Potter AA, Hanham AF, Nestmann ER (1985) A rapid method for the extraction and purification of DNA from human leukocytes. Cancer Lett 26:335–341

Raha S, Merante F, Proteau G, Reed JK (1990) Simultaneous isolation of total cellular RNA and DNA from tissue culture cells using phenol and lithium chloride. Genet Anal Tech Appl 7:173–177

Rahuel C, Dandieu S, Chausse AM, Mornet E, Rouger P, Cartron JP, Salmon C, Lucotte G (1983) Establishment of a DNA bank of human origin. Rev Fr Transfus Immunohematol 26:539–552

Ross KS, Haites NE, Kelly KF (1990) Repeated freezing and thawing of peripheral blood and DNA in suspension: effects on DNA yield and integrity. J Med Genet 27:569–570

Saez GT, Valls V, Muniz P, Perez-Broseta C, Iradi A, Oliva MR, Bannister JV, Bannister WH (1993) The role of glutathione in protection against DNA damage induced by rifamycin SV and copper(II) ions. Free Radic Res Commun 19:81–92

Sanger F, Nicklen S, Coulson AR (1977) DNA sequencing with chain-terminating inhibitors. Proc Natl Acad Sci USA 74:5463–5467

Scheibe O (1981) The meaning of specialist and adequate documentation for retrospective analysis of sicknesses (author's translation). Langenbecks Arch Chir 355:411–420

Scherczinger CA, Bourke MT, Ladd C, Lee HC (1997) DNA extraction from liquid blood using QIAamp. J Forensic Sci 42:893–896

Schneeberger C, Kury F, Larsen J, Speiser P, Zeillinger R (1992) A simple method for extraction of DNA from Guthrie cards. PCR Methods Appl 2:177–179

Schneider PM (1998) Recovery of high-molecular-weight DNA from blood and forensic specimens. Methods Mol Biol 98:1–7

Schreiber A, Amtmann E, Storch V, Sauer G (1988) The extraction of high-molecular-mass DNA from hair shafts. FEBS Lett 230:209–211

Schull WJ, Weiss KM (1980) Genetic epidemiology: four strategies. Epidemiol Rev 2:1–18

Schultz J, Rupprecht A, Song Z, Piskur J, Nordenskiold L, Lahajnar G (1994) A mechanochemical study of MgDNA fibers in ethanol-water solutions. Biophys J 66:810–819

Sharp PA, Sugden B, Sambrook J (1973) Detection of two restriction endonuclease activities in Haemophilus parainfluenzae using analytical agarose–ethidium bromide electrophoresis. Biochemistry 12:3055–3063

Sidow A, Wilson AC, Paabo S (1991) Bacterial DNA in Clarkia fossils. Philos Trans R Soc Lond B Biol Sci 333:429–432

Smith DA, Martinez AM, Ratliff RL (1970) Deproteinization with phenol of alternating polydeoxyadenylate-deoxythymidylate and other DNA-like polymers. Anal Biochem 38:85–89

Sokolov BP, Dzhemelinskii VV, Kalinin VN (1989) Isolation of high molecular weight eukaryotic DNA with the use of potassium acetate. Mol Gen Mikrobiol Virusol (6): 45–46

Soler C, Chardonnet Y, Guerin-Reverchon I, Chignol MC, Viac J, Euvrard S, Thivolet J (1992) Influence of fixation on human papillomavirus DNA detection in frozen and embedded paraffin lesions by in situ hybridization with biotinylated probes. Pathol Res Pract 188:1018–1027

Southern EM (1975) Detection of specific sequences among DNA fragments separated by gel electrophoresis. J Mol Biol 98:503–517

Steinberg KK, Sanderlin KC, Ou CY, Hannon WH, McQuillan GM, Sampson EJ (1997) DNA banking in epidemiologic studies. Epidemiol Rev 19:156–162

Sun F, Flanders WD, Yang Q, Khoury MJ (1998) A new method for estimating the risk ratio in studies using case-parental control design. Am J Epidemiol 148:902–909

Temin HM, Mizutani S (1970) RNA-dependent DNA polymerase in virions of Rous sarcoma virus. Nature 226:1211–1213

Thomas WK, Paabo S (1993) DNA sequences from old tissue remains. Methods Enzymol 224:406–419

Thomas WK, Paabo S, Villablanca FX, Wilson AC (1990) Spatial and temporal continuity of kangaroo rat populations shown by sequencing mitochondrial DNA from museum specimens. J Mol Evol 31:101–112

Thomson DM, Brown NN, Clague AE (1992) Routine use of hair root or buccal swab specimens for PCR analysis: advantages over using blood. Clin Chim Acta 207:169–174

Topic E, Gluhak J (1991) Isolation of restrictible DNA. Eur J Clin Chem Clin Biochem 29:327–330

van Aken WG (1986) Future trends in blood component preparation. Vox Sang 51 [Suppl 1]:67–71

van Schie RC, Wilson ME (1997) Saliva: a convenient source of DNA for analysis of bi-allelic polymorphisms of Fc gamma receptor IIA (CD32) and Fc gamma receptor IIIB (CD16). J Immunol Methods 208:91–101

Visvikis S, Schlenck A, Maurice M (1998) DNA extraction and stability for epidemiological studies. Clin Chem Lab Med 36:551–555

Vorlickova M, Khudyakov IY, Hejtmankova I, Kypr J (1991a) Circular dichroism studies of salt- and alcohol-induced conformational changes in cyanophage S-2L DNA which contains amino-2-adenine instead of adenine. J Biomol Struct Dyn 9:81–85

Vorlickova M, Sagi J, Hejtmankova I, Kypr J (1991b) Alkyl substituent in place of the thymine methyl group controls the A-X conformational bimorphism in poly(dA-dT). J Biomol Struct Dyn 9:571–578

Wadman M (1999) Human Genome Project aims to finish "working draft" next year (news). Nature 398:177

Walsh DJ, Corey AC, Cotton RW, Forman L, Herrin GL Jr, Word CJ, Garner DD (1992) Isolation of deoxyribonucleic acid (DNA) from saliva and forensic science samples containing saliva. J Forensic Sci 37:387–395

Watson JD, Crick FH (1974) Molecular structure of nucleic acids: a structure for deoxyribose nucleic acid. G.D. Watson and F.H.C. Crick. Published in Nature, number 4356 April 25, 1953. Nature 248:765

Weisberg EP, Giorda R, Trucco M, Lampasona V (1993) Lyophilization as a method to store samples of whole blood. Biotechniques 15:64–68

Weiss B, Richardson CC (1966) End-group labeling of nucleic acids by enzymatic phosphorylation. Cold Spring Harbor Symp Quant Biol 31:471–478

Weiss KM (1998) In search of human variation. Genome Res 8:691–697

Wilcockson J (1973) The use of sodium perchlorate in deproteinization during the preparation of nucleic acids. Biochem J 135:559–561

Wilcockson J (1975) The differential precipitation of nucleic acids and proteins from aqueous solutions by ethanol. Anal Biochem 66:64–68

Wilson RK (1999) How the worm was won. The C. elegans genome sequencing project. Trends Genet 15:51–58

Woodman R (1999) Wellcome Trust and drug giants fund gene marker database (in process citation). Br Med J 318:1093

Yang DY, Eng B, Waye JS, Dudar JC, Saunders SR (1998) Technical note: improved DNA extraction from ancient bones using silica-based spin columns. Am J Phys Anthropol 105:539–543

Yang Q, Atkinson M, Sun F, Sherman S, Khoury MJ (1997) The method of sib-pair linkage analysis in context of case-control design. Genet Epidemiol 14:939–944

Yates JR, Malcolm S, Read AP (1989) Guidelines for DNA banking. Report of the Clinical Genetics Society working party on DNA banking. J Med Genet 26:245–250

Yokota M, Tatsumi N, Tsuda I, Takubo T, Hiyoshi M (1998) DNA extraction from human urinary sediment. J Clin Lab Anal 12:88–91

Yokota M, Tatsumi N, Nathalang O, Yamada T, Tsuda I (1999) Effects of heparin on polymerase chain reaction for blood white cells (in process citation). J Clin Lab Anal 13:133–140

Zhang L, Cui X, Schmitt K, Hubert R, Navidi W, Arnheim N (1992) Whole genome amplification from a single cell: implications for genetic analysis. Proc Natl Acad Sci USA 89:5847–5851

Zimmerman SB, Little JW, Oshinsky CK, Gellert M (1967) Enzymatic joining of DNA strands: a novel reaction of diphosphopyridine nucleotide. Proc Natl Acad Sci USA 57:1841–1848

3 Microsatellite Genotyping

CHENI KWOK and KARIN SCHMITT

3.1 Introduction

Microsatellite genotyping has been widely used in the scientific community and has been the method of choice for many recent linkage analysis studies, especially genome scans designed to identify genes underlying common diseases (Ghosh et al. 1999; Risch et al. 1999; Xu et al. 1999). Microsatellite markers are simple sequence repeats consisting of repetitions of very short nucleotide motifs (usually 1–5 nucleotides) and can occur in perfect repetition, as interrupted repeats or together with another repeat type. Markers are categorized by the type of repeat motif and the most commonly used markers for human studies are dinucleotide repeats (often referred to as CA-repeats) which show a repeat pattern of two nucleotides. Other frequently used markers include tri- and tetranucleotide repeats. Microsatellite repeats (also known as simple sequence repeats) have been shown to be very abundant and highly polymorphic in many eukaryotic genomes (Litt and Luty 1989; Tautz 1989; Weber and May 1989). Simple sequence repeats are analyzed by polymerase chain reaction (PCR) amplification of a short genomic region containing the entire repeated sequence, followed by size determination of the repeat length by gel separation. All of the microsatellite markers described here for use in genotyping are easy to work with. More importantly for linkage studies, microsatellite markers have the highest level of informativeness amongst all available markers. New microsatellite markers are easy to identify by cloning directly from genomic DNA and are available abundantly throughout the genome. Accounting for their widespread use is the fact that microsatellite typing can easily be automated and exploited in a high-throughput fashion. Hence most genome scans for linkage studies have to date almost exclusively relied on microsatellite typing. A typical workflow for a high-throughput microsatellite genotyping laboratory is outlined in Table 3.1.

The abundance of microsatellite markers in many genomes, as well as the relatively easy identification of new microsatellite markers, have also resulted in the widespread use of these markers to build genetic maps for many organisms. This has led to projects that have systematically identified and mapped microsatellite markers in the human genome (Sheffield et al. 1995; Dib et al.

Principles and Practice
Molecular Genetic Epidemiology – A Laboratory Perspective
Ian N.M. Day (Ed.)
© Springer-Verlag Berlin Heidelberg 2002

Table 3.1. Genotyping flowchart

Sample preparation
DNA extraction
Determine DNA concentration and normalize
Prepare master and working stocks of DNA samples

Marker preparation
Test PCR conditions for all markers
Determine allele ranges for study population

Genome Scan
PCR all samples against all markers

Preparation for gel electrophoresis
Semiquantitative PCR product
Pool PCR samples based on relative concentrations and allele
    ranges
Add dye and size-standard before loading samples

Gel electrophoresis

Post-electrophoresis sample handling
Lane tracking
Size standard calling
Allele calling/allele binning
Check for Mendelian inheritance
Error detection
Prepare linkage files

1996). Having more than 10,000 polymorphic markers with exact mapping
locations made it possible to initiate genome scanning studies that required
dense marker coverage for all chromosomes. By investigating small chromosomal segments on a whole-genome basis, it became possible to start searching for linkages to markers in multigenic diseases. Since many study designs
require genotyping of more than 1,000 patient samples (about 400,000 PCRs
depending on desired marker density), the need for a cost-effective and high-
throughput genotyping system is apparent. The most popular systems involve
acrylamide gel electrophoresis of labelled PCR products, coupled with semiautomated allele calling and data analysis.

We will describe the complete genotyping process in this chapter, including
experimental setup and recommended bioinformatics and robotics systems, as
well as pointing out areas that require special attention when considering high-
throughput genotyping operations. We will also discuss improvements in the
current technology and describe future technologies that will make it possible
to genotype even larger study populations at increasing marker densities. The
last section of this chapter is devoted to a brief overview of possible applications for microsatellite genotyping.

3.2 Microsatellite Markers

Since the original report of Botstein and colleagues (Botstein et al. 1980), describing the construction of a human genetic linkage map using random single-copy DNA probes, extensive sets of genetic markers covering the whole genome have been generated (Donis-Keller et al. 1987; Sheffield et al. 1995; Dib et al. 1996; Wang et al. 1998). Currently, microsatellite markers are most commonly used for human genetic linkage studies, especially in the genetic dissection of complex traits. Microsatellite markers consist of repeat units containing two, three or four nucleotides (di-, tri-, or tetranucleotide repeats). They are very abundant in many genomes and dinucleotides are estimated to occur approximately once every 30 kb in the human genome (Litt and Luty 1989; Stallings et al. 1991; Beckman and Weber 1992). The number of repeat units can be subject to high mutation rates (Weber and Wong 1993) making microsatellite markers highly polymorphic in the human population. Since the total length of the repeat unit is typically less than 60 bp, it is possible to amplify such markers using primers flanking the repeat unit and analyzing the PCR products using short acrylamide electrophoresis runs. New microsatellite markers are easy to identify by hybridization of filter clones with a probe specific for a given repeat unit. Clones are sequenced to verify the number of repeats present as well as obtaining flanking sequence for primer design. Alternatively, sequence databases can be electronically screened for the presence of repeat motifs. This is the most useful approach when generating additional markers in a specific region of interest, such as a segment to contain a gene of interest. Repeat tracts exceeding a certain lengths (typically >20 bp) usually have a high likelihood to be polymorphic. Novel microsatellite markers are tested by PCR against a population sample to determine the degree of heterozygosity (i.e. the degree of informativeness for each marker).

Advantages of using microsatellite markers include their high heterozygosity values across all human populations, abundance in many genomes, easy and automatable assays by PCR followed by gel electrophoresis, and the availability of exact map locations on genetic, physical and radiation hybrid maps. Since dinucleotide repeat markers are more abundant than microsatellites with larger motifs, dinucleotide markers have been most commonly used. The advantage of tri- and tetranucleotide repeat markers is that the scoring of alleles following gel electrophoresis is expected to be easier since the patterns are less complex (alleles are more widely spaced) which presumably makes automation of allele scoring more robust. However, for a trained individual, there will be no difference in scoring either marker type and automated allele-calling software now exists to support genotyping using di-, tri- and tetranucleotide markers. Since the allele distributions for tri- and tetranucleotide markers have a broader range compared with dinucleotide markers, fewer markers can be combined to run together in the same gel lane (this pooling process is also referred to as multiplexing). This could potentially have a tremendous impact on the genotyping throughput (see discussion below).

The mutation rate of microsatellites is estimated to be between 10^{-2} and 10^{-5}, which for most markers is low enough to be still useful for linkage studies but is considered to be more variable within the class of tri- and tetranucleotide repeats (Weber and Wong 1993). Markers that show a mutation rate greater than 10^{-2} should not be used in a linkage analysis study. In summary, we cannot make any strong recommendations for dinucleotide markers vs. tri/tetranucleotide markers since there are many factors that have to be considered (ease of scoring, degree of multiplexing, mutation rates and availability of whole genome marker panels; see below). Other attractive alternative markers for genotyping are available. Besides microsatellite markers, the most common type of genetic variation in the human genome is the single nucleotide polymorphism (SNP). The use of SNP markers for high-throughput genotyping is described elsewhere in this volume.

Public efforts have generated large numbers of microsatellite markers as well as their exact genetic map locations and allele distributions (CHLC, Genethon). This information was immediately adopted to group together sets of evenly spaced markers spanning the whole genome that could be used for localizing a putative gene to a particular region in the genome (Levitt et al. 1994; Reed et al. 1994; Lindqvist et al. 1996; Yuan et al. 1997). Currently, the two most commonly used microsatellite marker sets include the ABI PRISM Linkage mapping set from PE Applied Biosystems and the CHLC Human Screening Set/Weber distributed by Research Genetics. The ABI PRISM linkage mapping set consists of fluorescently labeled PCR primer pairs selected to amplify highly informative dinucleotide markers taken from the Genethon human linkage map (Weissenbach et al. 1992; Gyapay et al. 1994; Dib et al. 1996). The primers have been redesigned and optimized to run under similar PCR conditions for maximum throughput. Three different mapping sets are available for 20, 10 and 5 cM coverage of the human genome. The CHLC Human Screening Set/Weber is distributed by Research Genetics. This set is primarily based on tri- and tetranucleotide markers chosen for good genome coverage, ease of scoring and high heterozygosity. Both a high density (10 cM) and a low density (25 cM) set are available. Other laboratories have put together their own marker sets that can be more specific for a given population and contain markers that are optimized for their specific equipment. Although the commercially available marker sets have been screened for robust performance, individual laboratories still might experience some difficulties in getting all markers to work.

Some fine-tuning of PCR parameters and occasional substitutions of particularly difficult markers are commonplace. The number of markers actually required for a study will depend on the specific application. Factors including the strengths of a signal to be detected, the number of genes involved and the number of affected families/samples available are all important. Possible applications that will dictate marker density include genotyping of isolated populations, linkage disequilibrium/association studies, polygenic disorders, affected sib-pair and other family-based studies, exclusion mapping, monogenic disorders, loss-of-heterozygosity mapping and mapping in specific chromosomal areas of interest. In general, the two most important factors that determine success are obtaining high marker heterozygosities with even genome coverage and the available patient sample size. It is usually easier (and less expensive) to genotype additional markers instead of increasing the sample size by collecting new patient material.

Microsatellite markers can be grouped into sets based on allele ranges, similar PCR conditions, or according to their chromosomal location (chromosome-specific sets). Usually, sets are not multiplexed during the PCR step but are combined following PCR amplification prior to electrophoresis. Multiplexing several markers during PCR is possible, but it is often difficult to achieve good amplification across all markers in a single PCR reaction. In order to ensure success, additional primer redesign and careful PCR optimization are required. Other laboratories (Shuber et al. 1995; Lin et al. 1996; Wang et al. 1998) have suggested the addition of common tails to all primers, trying to achieve similar melting temperatures for all primers, as well as keeping the size of the PCR product to a minimum. The degree of multiplexing (i.e. the total number of marker sets in a single lane) for a typical genome scan has a tremendous effect on the throughput. Since gel electrophoresis is considered the most time-consuming aspect of a study, pooling of PCR products could double the throughput when comparing pooling six markers versus pooling twelve markers per lane. When using publicly available sets, the degree of multiplexing is usually fixed and determined by the number of markers in a given set. For sets put together individually, we recommend to aim for multiplexing on average about 8–10 markers per set. This number represents a good balance between achieving a high-throughput and making sure that allele ranges from two similarly sized markers will not be overlapping.

Additional work will be required to determine allele ranges for new markers in order to group markers together in a set. In our laboratory, we routinely determine allele ranges for a given marker against the sample population to be studied or using the CEPH reference pedigrees. We use four pools of 12 unrelated (and unaffected) individuals plus 4–6 separate samples. This also allows us to judge the morphology of the alleles and obtain data on allele ranges in special populations. Based on this information, we put together our own marker sets for genome scanning or fine mapping. When choosing markers for putting together individual sets, the degree of informativeness for a given microsatellite marker (also called heterozygosity) should be one of the most important considerations. The heterozygosity of a microsatellite marker is calculated from the allele frequencies in the population. The publicly available panels have average heterozygosities of 0.7–0.8 and have been optimized in that respect. However, one has to be aware that most of the data have been obtained from Caucasian samples. Actual heterozygosity values, especially in a more isolated population, can differ drastically. Having an abundance of markers with low heterozygosity values in a genome scan is inefficient but can be compensated by using additional markers or by substituting markers to achieve better coverage. In addition, it is important to have the exact map location for any marker used in the studies described here. The publicly available marker sets only include markers whose exact genetic map location is known. Other sources of mapping information can be obtained from the literature on genome maps (Sheffield et al. 1995; Dib et al. 1996).

3.3 The Genotyping Process – Experimental Considerations

The genotyping process usually starts with the DNA sample preparation. It is important that the DNA is of the highest quality since this will have a great impact on generating robust and easily callable genotypes. DNA extraction can be performed using commercially available kits or following published protocols. We have found that some extraction protocols give excellent yields but the DNA might not always perform well during PCR. When deciding which DNA extraction method to use, it is recommended to test a range of samples for their PCR performance. Often DNA samples are already available but DNA quality can deteriorate during long storage times. In order to increase sample yield from such deteriorated samples, we have successfully used several of the pre-amplification protocols to increase the number of genotypes per DNA sample and to achieve superior results with degraded DNA (Telenius et al. 1992; Zhang et al. 1992).

Following DNA extraction, the DNA has to be quantitated using spectrophotometric or fluorometric methods. We routinely prepare stock DNA of normalized concentrations (usually $10 \times$) and a working stock for daily usage. All samples are bar-coded and recorded in a computer database. We keep information on DNA concentrations, arrival of the sample in the laboratory, sample ID, study ID, any unusual appearance of the samples and what experiments the samples have been used for. The most widely used format for genotyping is the 96-well microtiter plate (or 192- and 384-well plates). When arranging DNA samples on a given plate, we routinely include several positive controls (CEPH individuals). We also make sure that DNA samples belonging to the same family are not run in adjacent gel lanes to minimize genotyping errors. Before starting a whole genome scan, we test all DNA samples against two sets of highly polymorphic markers. This assures not only that the DNA quality/PCR performance is acceptable, but also helps to rule out any "paternity" errors in a given family. We will try to resolve such errors (often resulting from a simple sample switch or data entry error), take out or replace the sample, or change the family structure accordingly.

Independent of using commercially available marker sets or individual sets, we recommend (re-)optimizing all PCR primers. This process can include testing several annealing temperatures, Mg^{2+} concentrations as well as primer concentrations to fine-tune PCR conditions for an individual laboratory. These parameters can influence the yield as well as the appearance of the genotypes/peaks (prevalence of stutter bands). We try to select primers that work under identical conditions to minimize the time required to set up a genome scan. In cases of poor PCR performance or other difficulties such as null-alleles (see Sect. 3.4.4.4), redesigning the primers for the same locus can usually solve the problem. Protocols such as touchdown (Don et al. 1991) and hotstart PCR give superior genotyping results. There are also several commercially available enzymes (for example, AmpliTaq Gold distributed by PE Applied Biosystems or Platinum GenoTYPE Taq DNA Polymerase from Life Technologies) that result in the same effects as those observed using touchdown/hotstart PCR but without the additional experimental steps involved.

Two additional factors that determine how easy it will be to score a given microsatellite genotype are stutter bands preceding the primary peak (probably caused by slippage during amplification of dinucleotide repeats) and "Plus A" (nontemplated addition of an extra adenine nucleotide at the 3′ end of the PCR product). Stutter bands can be minimized by choosing the appropriate PCR conditions (see above) and can even be incorporated in the allele-calling process to achieve more accurate allele discrimination. In contrast, "Plus A" addition can easily lead to miscalled genotypes, especially when the process is only affecting a subset of PCR products (incomplete addition) or changes over time/occurs intermittently. The appearance of "Plus A" seems to be primer-specific, influenced by the sequence at the 3′ end of the PCR product as well as the PCR conditions. There are several solutions reported in the literature and we routinely follow the recommendations for cycling profiles/holding times as well as primer modifications (Smith et al. 1995; Brownstein et al. 1996; Ginot et al. 1996; Magnuson et al. 1996).

When working in a high-throughput environment, it is not only important to generate a large number of genotypes but also to ensure that the data generated is of the highest quality. In order to measure the genotyping error rate (reproducibility), a certain number of genotypes should be performed in duplicate. Genotyping error rates calculated from such experiments should be less than 1%. Additional procedures can be put in place for high quality assurance of all DNA samples and performance of PCR primers (see recommendations above). Although some PCR failures are commonplace, the average PCR success rate can be expected to be around 92–98% for any particular marker. Primers that do not perform at that level should be replaced. In addition, we routinely test all batches of new reagents on small samples to ensure high quality before setting up large experiments. PCR contamination is usually not much of a problem when following standard procedures to avoid contamination (Kwok and Higuchi 1989).

3.4 Data Analysis

With the recent advancements in high-throughput genotyping technologies, acquisition of genotyping data is becoming increasingly faster as well as cheaper. However, with the generation of large volumes of raw data, conceptually simple tasks such as lane tracking, allele sizing and binning will impose bottlenecks on throughput capacity. In this section, various approaches to the automation of the data analysis steps required for fluorescently labeled microsatellite typing as well as software available to aid in data processing will be discussed. The approaches described here are biased towards using the DNA sequencers for fragment analysis available from PE Applied Biosystem. However, the principle steps involved in data analysis should be applicable to most of the slab-gel electrophoresis equipment currently available commercially. In addition, in any high-throughput setup, it is crucial to ensure that the accuracy and quality of the genotypes generated is not being compromised. Several approaches to minimize genotyping errors as well as error-checking mechanisms available to date are reviewed here.

3.4.1 Lane Tracking

The first manual step for processing the gel file for gel-based electrophoresis platforms is lane tracking. This is a process to demarcate lanes and lane boundaries by locating the strongest fluorescent signal for each lane. It is crucial to ensure that the lane-tracking process is performed accurately. In addition, it is important to make sure that all expected size-standard fragments are captured during lane tracking. Inaccurate tracking may result in scoring of size-standard bands from adjacent lanes, leading to genotyping errors that can be difficult to resolve during later stages of data processing. Although in theory the lanes should run in a straight line, due to various experimental factors (including heterogeneity in gel composition, changes in gel thickness, uneven heat distribution, fluctuations in the power supply, variable amounts of salt present in the reaction, buffer leaks and background noise interfering with detection), the curving of lanes and other gel artifacts are frequently observed. In addition, poorly formed wells and unequal loading will result in considerable variability in lane width, lateral spacing and signal intensity. These nonlinear lanes, background stripes or patches on gel images as well as other imperfections pose a challenge for automated lane tracking. The tracking of each lane in order to follow the nonlinear running of the fragments is a simple but very time-consuming manual step. Various algorithms, including neural net trackers (Golden et al. 1993), have been developed to facilitate automatic tracking of sample lanes. In addition, computer programs based on novel curve fitting methodologies or quantitative gel imaging have been adopted to aid lane tracking (Perlin et al. 1995; Cooper et al. 1996; Shadle et al. 1997). Experimental implementations such as staggered loading or running a denser set of molecular weight standards aid the lane-tracking process. With the advent of capillary sequencers and other non-gel-based platforms for accurate DNA fragment sizing (see Sect. 3.8), lane tracking will not be an issue in the near future.

3.4.2 Allele Sizing and Size Standards

The precise and accurate size determination of DNA fragments in microsatellite typing is the most crucial step in the genotyping process. Accurate sizing of alleles will facilitate allele binning (see Sect. 3.4.3). Sizing of DNA fragments is achieved by fitting a calibration curve based on mobility data from size standards. For automated DNA sequencers, the sizing of the alleles is measured by comparing against the co-migration of a labeled standard molecular weight ladder in each lane (Mayrand et al. 1992). Various calibration methods are available, and it is important to use the same method of calibration throughout one study, as different calibration methods affect the precision of allele sizing (Ghosh et al. 1997).

Furthermore, the choice of the size standards used will affect the accuracy of the allele sizing. The spacing of the standard fragments is an important consideration for accurate typing. As a general rule, we recommend the bands in the size standards to have even band intensities and the fragments cover the entire size range of alleles to be measured. In addition, the bands should be evenly spaced, and the spacing between fragments should not be wider than 25 bp. The average standard deviation is shown to be less than 0.2 bp when using standard size ladders of approximately 20-bp spacing (M.J. Daly, M.I.T. Whitehead, pers. comm.). It is important to predetermine the allele ranges expected for a given population, and choose the appropriate molecular weight ladders for sizing. However, it is crucial to use the same size standard throughout the entire study, to ensure data compatibility. Commercially available size standards include the GeneScan size standards developed by PE Applied Biosystems (GeneScan 350 containing 12 fragments between 35 and 350 bp; GeneScan 500 consisting of 16 fragments between 35 and 500 bp; GeneScan 1000 has 17 fragments ranging from 47–946 bp). Other fluorescently labeled size standard markers on the market include the ALFexpress sizers 50–500 (Amersham Pharmacia Biotech) consisting of 10 fragments with 50-bp increments from 50–500 bp; MapMarker (BioVentures, Inc.) with average spacings of 10–20 bp for sizing up to 400 bp and two IRDye sizing standards developed by Li-Cor, Lincoln (50–350 sizing standard contains 14 bands from 50–350 bp and 50–700 sizing standard has 18 fragments with approximately 50-bp spacing from 50–700 bp). Several investigators have custom made size standards to cater to their particular marker sets and to reduce cost (G. Gyapay, pers. comm.). In addition, many laboratories have constructed additional external size standards specific to the loci under investigation to correct for gel-to-gel variation (see detailed discussion in Sect. 3.4.4.3). These external standards can be made using DNA from a mixture of genomic DNA or using reference individuals of the CEPH family 1331 (parents) available from the NIGMS Human Genetic Mutant Cell Repository (Camden, New Jersey, USA).

3.4.3 Allele Calling and Allele Binning

The data obtained after automated fragment sizing are accurate to two decimal places. There is usually a spread of ±0.5 bp around the actual base-pair size DNA fragment size due to allelic drift (Sect. 3.4.4). Allele calling is a process that converts imprecise DNA fragment sizes to discrete alleles that these size calls represent. The assignment of alleles is straightforward for the trained eye. However, as there are a large number of genotypes to be processed, it is highly desirable to automate the allele-calling process. The general principle for automating the allele-calling process is to define the tolerance or stringency level to represent the minimum distance within which the alleles must lie. Various algorithms have been developed for facilitating the automated calling. For example, automated binning and adjustment software (ABAS) (Ghosh et al. 1997); genetic typing software (Molecular Dynamics, Sunnyvale); TrueAllele (Cybergenetics, Pittsburgh); Genotyper (Applied Biosystems); Scorer (Marshfield Centre for Medical Genetics); Gene ImagIR software (Li-Cor, Lincoln); and StatGene software developed by the Whitehead Institute for Biomedical Research (www.biocomputing.fi). It is, however, still mandatory to perform a manual inspection to ensure accurate allele calling after running automated allele callers. Hence, allele calling remains a major bottleneck in the genotyping process. Some of the software, such as TrueAllele (Cybergenetics, Pittsburgh) and the automated genotyping programs by Idury and colleagues (Idury and Cardon 1997) have a built-in facility to provide a 'quality measurement' for the allele calling. This provides a mechanism to allow allele callers to prioritize the manual inspection process. Allele binning is the process of converting alleles from real-valued DNA fragment sizes into discrete segregating units (or bins). The assignment of a 'bin label' is a labor-intensive process, and it is highly desirable to bin the entire data set simultaneously in order to define the bin ranges accurately. Software products such as ABAS (Ghosh et al. 1997) and StatGene (www.biocomputing.fi) have incorporated an automated allele-binning facility in their genotyping packages.

3.4.4 Allelic Dispersion

A major source of error in genotyping is due to the difficulty of accurate sizing for a given allele as a result of allelic dispersion or allelic drift. The various causes of allelic dispersion and different experimental approaches and algorithms to correct these errors are discussed below.

3.4.4.1 Allele Plus A

Allele Plus A is caused by the nontemplated addition of adenine by Taq DNA polymerase. This addition of a single base is shown to be template dependent, i.e. marker specific (Smith et al. 1995; Magnuson et al. 1996). Furthermore, the extent of Plus A addition for a given allele is not consistent, and variables contributing to marker specificity of the Plus A effect are not well understood. As a result of this addition of A, spurious noise peaks of one nucleotide greater than the true allele size are generated. It is thus not surprising that automated allele calling will fail when a ladder of bands spaced one nucleotide apart is generated with partial Plus A modification in addition to the 2-bp shadow bands produced by the stuttering effect (see below). Various experimental protocols have been designed to overcome the allele Plus A problem. These include eliminating extra bases by enzymatic methods (Ginot et al. 1996), modification of PCR cycling profiles and PCR conditions (Smith et al. 1995; Magnuson et al. 1996), as well as modifications of primer sequences such as PIG-tailing that drive the completion of adenylation of PCR products (Brownstein et al. 1996). All these approaches have been shown to increase the accuracy of allele sizing. Furthermore, algorithms based on subtracting 1 bp from the size measured for the markers prone to the Plus A phenomenon have been developed (Perlin et al. 1995).

3.4.4.2 Stuttering Artifact

Due to the tandem repeat units present for a given microsatellite locus, PCR stutter bands are often observed as an experimental artifact. These shadow peaks have the characteristic of diminishing amplitudes that are 2, 4, 6 bases shorter than the true dinucleotide repeat allele. The stuttering artifact is probably due to the slippage of DNA polymerase or slipped strand mispairing (Hauge and Litt 1993). The presence of these stutter peaks severely hampers automatic allele calling when a composite pattern of overlapping stutter bands occurs between two closely spaced alleles of a heterozygous individual. Several computer algorithms have been developed to overcome this and even to make use of stutter patterns to achieve more accurate allele calls. For example, algorithms have been developed to exploit the reproducibility of stutter bands unique to each marker, thus removing the stutter artifact mathematically via deconvolution methods (Perlin et al. 1995). Other models rely on a quantitative analysis of relative intensities contributed by stutter peaks in order to resolve overlapping stutter patterns (Miller and Yuan 1997).

3.4.4.3 Gel-To-Gel Variability

In any high-throughput setup, a number of DNA sequencers are used to generate data for a particular study. It has been noticed by various groups that the variability between different runs on the same machine and runs performed on different machines is significant enough to cause inappropriate allele binning for adjacent alleles (Ghosh et al. 1997; Idury and Cardon 1997). In order to address the issue of allelic drift, various controls were incorporated into the experimental setup in order to perform a normalization procedure to correct for gel-to-gel variability. For instance, in addition to running internal size standards in every lane to calibrate lane-to-lane variation, various groups have incorporated controls to allow external calibration of machines. For example, in the FUSION study, control DNA samples from Centre d'Etude du Polymorphisme Humain (CEPH) were run concurrently against the markers under study to perform external adjustments to different runs as well as to increase the accuracy of fragment sizing (Ghosh et al. 1997). Others have adopted strategies to incorporate external size standards, i.e. DNA fragments of known sizes in one or more lanes in the experimental setup (Mansfield et al. 1994). Various computer algorithms include an automated adjustment algorithm to allow for allelic drift; for example, automated binning and adjustment software (ABAS) (Ghosh et al. 1997), a software program developed by Idury and colleagues (Idury and Cardon 1997) and software for normalization using external standards (Mansfield et al. 1994)

3.4.4.4 Null Alleles

Null alleles or allele nonamplification occurs when a given allele is not represented at a given locus. Null alleles are one of the sources of genotyping errors, resulting in apparent non-Mendelian inheritance, and also decreasing the power of a given study. It has been shown that, in addition to the variable number of repeat units for a given locus, regions directly flanking the CA repeat can also be variable. As a result of this variability, primers flanking a microsatellite locus fail to bind at the priming sequence for the PCR reaction (Callen et al. 1993; Koorey et al. 1993; Grimaldi and Crouau-Roy 1997). In order to reduce the incidence of allele nonamplification, care has to be taken to select suitable flanking regions for primer binding or to redesign primers for problematic markers.

3.4.5 Approaches To Minimize Errors During Genotyping

With the increasing sophistication of algorithms for automated allele calling, as well as the improvements in platform technologies and reagents for performing genotyping, there is a steady rise in throughput as the various bottlenecks in the genotyping process are being addressed. However, as the amount of data generated accumulates, it becomes necessary to build in quality-control checkpoints as well as error-checking mechanisms to ensure data accuracy and to minimize errors in genotyping (Buetow 1991). Marker typing errors in a linkage data set will either negate a true recombination or introduce spurious crossovers. In building genetic maps, errors in estimation of recombination fractions result in inflation of genetic map length, and a lowering of the confidence of the correct genetic order (Lincoln and Lander 1992; Goldstein et al. 1997). For linkage studies, the undetected errors will result in a loss of statistical power. Estimates of errors in some existing genotype databases are in the range of 0.5–1.5% (Buetow 1991; Lasher et al. 1991; Brzustowicz et al. 1993). Implementation of various quality-control measures in the workflow will minimize experimental errors during genotyping. It is also good practice to obtain a realistic measurement of the genotyping errors in a high-throughput setup (see detailed discussion in Sect. 3.3). Furthermore, with the rise in throughput, it will be increasingly important to incorporate various error-checking mechanisms following allele calling to check for data accuracy and also to identify the source of potential errors (see below).

3.4.5.1 Checking Mendelian Inheritance

In a linkage data set, incompatibilities with Mendelian inheritance may result from incorrect marker typing, errors in data entry, pedigree errors such as false paternity, unknown adoption or sample switches. Confirmation of Mendelian consistency is routinely used as an initial screen for error checking of linkage data. In order to handle large amounts of data, and also to save time in tracing the sources of non-Mendelian inheritance, various computer programs have been developed to flag these inconsistencies prior to further linkage analysis. For example, MENDEL, a software package for pedigree analysis, was developed to perform iterative filtering of nonallelic fragments (Stringham et al. 1996). Allele filtering, i.e. discarding alleles present in an individual but absent from both parents, as well as genotype filtering, i.e. checking for Mendelian consistency, are checks incorporated into MENDEL. This is especially useful in providing the automatic identification of a small number of individuals most likely to be the source of inconsistency in the pedigree. Another program that detects marker genotype incompatibilities is UNKNOWN (a preprocessing program from the LINKAGE and FASTLINK programs; Cottingham et al. 1993). By inferring possible genotypes and mating combinations for each untyped person in the pedigree using Boolean logic, UNKNOWN attempts to identify in which of the nuclear families the incompatibilities arise. Several commercially available software packages have also been developed to check for inconsistencies with Mendel's rules of inheritance, for example, Genotyper (Applied Biosystems, Foster City) and Gene ImagIR software (Li-Cor, Lincoln).

3.4.5.2 Approaches To Identify Marker Typing Errors Consistent with Mendelian Inheritance

Although Mendelian checks provide a preliminary screen for errors, it has been noted that most typing errors do not result from discordances with Mendelian inheritance (Lathrop et al. 1983; Dracopoli et al. 1991). Although the ideal way to reduce errors is to perform duplicate typing, this is neither feasible nor cost-effective for large-scale studies. There are various approaches that have been adopted to address the issue of marker typing errors consistent with Mendelian inheritance. One approach is to identify multiple recombination events within a short genetic interval (intralocus recombinants) or between small physical intervals (intracluster recombinants). Since these are statistically unlikely events under the genetic model, such apparent recombinations may point to genotyping errors. The CHROMPIC option of the CRIMAP program (Green et al. 1988) has built-in checks to screen for double crossing-over events in a small interval. In addition, map construction software such as MultiMap (Matise et al. 1994) and the CHLC software (Buetow et al. 1994) have integrated error-detection utilities for identification of apparent recombinants between closely linked markers. Several statistical approaches have also been devised to aid error detection and correction. One error-checking approach is to calculate the relative (posterior) probabilities for the true relationships in each family (Lathrop et al. 1983). By adapting the likelihood ratio test, GENOCHECK software was designed to pinpoint any unlikely genotype error (Ehm et al. 1995, 1996; Brzustowicz et al. 1993). Other software such as LINKAGE/FASTLINK (Cottingham et al. 1993) and MENDEL (Lange et al. 1998) have built-in error-detection methods to identify typing errors.

3.5 Data Management

As the throughput increases, it is crucial to have a robust and efficient data management system to perform sample tracking, workflow management and data management. Depending on the throughput and budget constraints, off-the-shelf software such as Filemaker Pro and Microsoft Excel can be used to organize various sample information, details of marker panels as well as plate organization and other important information. Project management tools such as Microsoft Project can be used to schedule the workflow and track the progress of different projects. For processing and storage of genotyping data, most medium-throughput laboratories make use of software provided by the various manufacturers. For higher-throughput facilities, it is commonplace for such laboratories to process a large number of DNA samples (tens of thousands) using a variety of marker sets, and to generate millions of genotypes a year. It is an enormous challenge to organize sample storage, to keep track of all samples being processed and to organize and store the sheer amount of data without a laboratory information management system (LIMS). In general, LIMS should provide centralized data warehousing, thus ensuring data integrity and facilitating data retrieval. In addition, the LIMS can be designed to aid in workflow management and to generate reports. Furthermore, an LIMS can be built to facilitate automated data processing and to track data through the data-processing pipeline.

3.5.1 Sample Registration and Laboratory Workflow Management

LIMS can be viewed as a central data depository, as well as a management tool for sample and workflow tracking. For instance, the LIMS may be built to allow the input of all available information regarding the samples, marker sets, working plates and gel runs, etc. Individual samples can be bar-coded to facilitate the tracking process. Although this seems to be a time-consuming step, it allows the accurate recording of all information pertaining to the samples, and also provides flexibility in building up large numbers of sample plates. In addition, the LIMS can aid in experimental planning by generating sample layouts, organizing panels of primers, creating barcodes for samples, working plates and gels, generating the loading order of gels, etc. These tools can help to increase throughput, reduce costs, and improve overall efficiency. Furthermore, the LIMS can be built to generate reports for each stage of the workflow (such as work-turnaround summaries) as well as to facilitate troubleshooting. It can also provide reports of sample locations and checkpoints for various quality-control steps throughout the process, thus improving laboratory efficiency, while maintaining the quality of the data produced.

The LIMS can also be designed to reduce the amount of manual data transfer from one stage of the workflow to the next and to facilitate automation of laboratory processes. For instance, prior to the loading of pooled samples onto the electrophoresis instrument, various parameters need to be keyed into the machine. With the electrophoresis machines developed by PE Applied Biosystems, sample sheets containing information about the markers, samples, loading order and electrophoresis conditions require manual entry prior to electrophoresis. This time-consuming step can be circumvented by generating sample sheets automatically using LIMS. There is software available to aid in sample and workflow management. For instance, SQL*GT (Perkin Elmer Informatics, Foster City) is a sample and project management software that can track samples, manage workflow and generate sample sheets. Other software such as Gene ImagIR (Li-Cor, Lincoln) also allows tracking of family relationships, affection status, liability classes, etc., in addition to organization of workflow in the laboratory.

LIMS can also serve as the central controller for various automated data-processing functions. The integration of data collection and automated data analysis can prove to be a challenge when a choice of different DNA electrophoresis equipment from various manufacturers, for example, PE Applied

Biosystems Inc (Foster City, CA), Li-Cor (Lincoln, NB), and others is employed. As every type of instrumentation has a unique suite of data-analysis tools, and analyses are performed on the data collection instrument itself, there is a need to transfer data from the data collection machines, and also to perform external adjustments to ensure data integrity. In order to increase processing efficiency and to perform batch data analysis from various instruments, many investigators have removed the analysis components from the machines, thus separating the data-collection process from the data-analysis process (Golden et al. 1993; Hall et al. 1996). The LIMS can be designed to pass data from the electrophoresis machine through different stages of the data-processing steps (see Sect. 3.4) and to store the finished genotypes. All finished genotypes can be exported from the LIMS for various statistical analyses in the appropriate customized formats. Several off-the-shelf commercial LIMSs are available to facilitate data-flow management and/or centralize data storage for finished genotypes. For instance, BioLims (Perkin Elmer Informatics, Foster City) provides automated data collection. Discovery Manager (Genomica, Boulder) is a software package which additionally provides population genetics tools such as pedigree management to examine the distribution of a disease within a family, tools for patient record management, and storage of associated phenotypes and genotypes. A commercial high-throughput genotyping setup is usually organized as a core facility, providing service genotyping to different groups. A typical genotyping facility is required to concurrently process different types of studies, (see Sect. 3.8) which may need different experimental designs and workflows. As a result, most setups have a custom-built LIMS to cater to their various requirements and allow for adaptation to changes in platform technologies (see Sect. 3.7). Companies such as Cimarron Software, Salt Lake City, aim to provide and support such customized LIMSs.

3.6 Hardware

3.6.1 Robotics

There are several potential rate-limiting steps in typing microsatellite markers for large-scale mapping projects such as whole genome linkage analysis. The first in the overall workflow is setting up thousands of PCR reactions. This can be carried out using a 96- or 384-well format and multichannel pipetters, but robotic equipment will provide a suitable alternative. Some immediate benefits will include greater consistency of pipetting, avoidance of sample mix-ups, and potential savings of time and therefore cost. Many laboratories will want to consider the purchase of robotics equipment to support not only PCR setups but also multiplexing of markers following PCR amplification, addition of loading dye and preparation of various PCR cocktails. However, when considering the cost for the purchase and maintenance of robotic equipment, one only starts to save money when operating large-scale setups. Otherwise, there will be no substantial personnel/time savings, although the pipetting error rate will be greatly decreased when making use of robotics equipment. For most laboratories, it will not be necessary to consider buying a complete setup that would automate every step in the genotyping process with minimal human intervention (from PCR setup to gel loading). Instead, for a medium-size operation, it might be best to divide the process into several steps and only automate selected steps. One example would be to automate the pipetting of DNA samples or just the multiplexing step. For example, a typical whole genome scan usually involves a fixed number of DNA samples that can be prepared in advance and diluted to the required PCR concentrations. Plates can be aliquoted in advance using a simple 96-pipetting robot including only input and output trays. By preparing all required plates in advance, contamination between DNA samples can be minimized since pipetting probes only have to be cleaned when switching to different DNA stocks. These plates can be stored at 4 °C for several days to weeks (or dried down for easy storage) and PCR cocktails can be added manually at a later time.

For any size laboratory, additional benefits could be obtained by automating gel loading. While it has proven difficult to design reliable automated loaders that would simply attach to the instrument and load all samples without human intervention, another alternative has recently been described (Erfle et al. 1997; Cassel and Guttman 1998; Guttman 1999). This system uses membrane combs that can be spotted with the sample robotically or on the benchtop. The samples are loaded onto the teeth of such combs and the comb is subsequently inserted directly above the edge of the gel. Using this system can save considerable amounts of time and loaded membranes can be stored for later use or even be reloaded to save cost. Paper combs are currently marketed by Licor (www.licor.com) and The Gel Company (www.gelcompany.com).

Many companies offer robotic equipment; Table 3.2 provides a summary of several models. This summary is not meant to be all encompassing but gives a snapshot of the models that are currently widely used in the genotyping and sequencing community. The table also lists some of the factors that should be considered when comparing different models. Each robot is a relatively simple pipetting station that can handle 96- and 384-well plates with high precision. Critical determinants when choosing among the different robotic instruments are the smallest reliable pipetting volume (for reducing PCR volumes and giving associated cost savings), pipetting accuracy at low-volume, and the number of plates that can be processed without manual intervention. Other factors include cost of the instrument and required disposable items, ease of use (software), and additional features such as cooling platforms, stackers and expandability. One of the most important factors is whether the process requires on-line washing of the pipetting probes or removable tips. This is almost certainly necessary and critical when pipetting many different DNA samples while setting up PCRs, since even small amounts of contamination can

Table 3.2. Commercially available robotic equipment

	Jenoptik Bioin-struments GmbH JOBI series	Hamilton Microlab 4200	Robbins Scientific, Corp. Hydra	Tecan Genesis series	Beckman Multimek 96	TomTek Quadra 96SV	CCS Packard Plate Trak
Basic design	96-Well pipette tips including stacker or turntable; washing station optional	Multipurpose probes or tips (4,6,8,12) or other heads; interchangeable deck space including washing station	96-Well microdispenser including plate positioner or 100 plate robotic plate handler (Twister)	Robotic sample processor for multipurpose liquid handling includes flexible worktable and 4- or 8-channel probes	96-Channel pipette using disposable tips with 6 stages to hold plates; can be expanded with stacker carousel for 2 × 45 plates	96-Channel pipette with 6 stages to hold plates; exchangeable plastic tips or 96-probe fixed head; dual stacker upgrade to 50 plate capacity available	Robotic liquid handling system includes bi-directional, linear conveyor system with modular design including wash station
Pipetting volume	0.5–100µl depending on pipetting head	1→100µl depending on syringe size	0.1–100µl with 100-µl syringe	0.5–10µl with low-volume option	1–50µl with 50µl tips (200-µl tips also available)	0.5–60µl with 96-probe head (other heads available)	5–750µl with 96 fixed tip heads (other tip heads available)
Pipetting details	96-Well plate with 96-head; manual exchange of tip magazine	96- or 384-Well plates	96- or 384-Well plates	96- or 384-Well plates	96- or 384-Well plates; different pipetting heads available	96- or 384-Well plates	96- or 384-Well plates
Accuracy	1 µl ± 7%	1 µl ± 3% for 10-µl syringe	1 µl ± 3% for 100-µl syringe	for <5µl ± 10%	not available	<3% variation	<1% at 50µl
Workspace	Turntable for 8 plates plus washing station; dual stacker for 2 × 50 plates plus washing station	Modular workspace includes space for 34 plates plus wash station	Source and target plate including syringe washing station for Hydra-PP model; 100 plates for Twister set-up	Several worktable sizes from 12- to 33-plate capacity including wash station	6 Plates or 2 × 45 plates with stacker; washing station available	6 Plates plus wash station	Each stacker holds 50 plates (expansion to >4 stackers possible)
Cooling platform	No	Custom-designed	No	Optional	No	No	No
PCR applications	Well suited for medium setups	Well suited for high-throughput	Accurate dispensing, includes manual plate handling; Twister model allows extended walk-away operation	Well suited for high throughput	Robust system offering flexibility	Well suited for medium setups with easy expansion	Good model for any setup

lead to genotyping errors. Washing stations are usually the preferred alternative since expenses for disposable pipette tips can be substantial. In addition, one has to consider whether the same robot should be used for setting-up PCRs as well as post-PCR sample handling (multiplexing and preparation for gel loading). It would be best to have two separate robots available to avoid cross-contamination.

The robots available from Hamilton and Tecan offer large workspaces that can be occupied by microtiter plates or other parts (such as washing stations or custom-designed racks to hold other types of tubes). They offer the greatest flexibility but have constraints on the number of microtiter plates that can be set up in a single run. However, this limitation can be overcome by integrating a robotic arm and stackers to store additional plates. Other robots (Jenoptik and CCS Packard) are equipped with stackers that feed large numbers of plates into a pipetting station. More basic models include the Hydra, Tomtek and Multimek, that in their minimal configurations include small pipetting stations that can be expanded (even at a later stage) by stackers or by adding twister arms. These provide a good choice for a small genotyping operation that can be expected to grow over time. In addition, the robots listed can be distinguished by their pipetting heads, that can have easily exchangeable tips, fixed tips or fixed probes. Again, a choice between these different options is dictated by which laboratory operations will be supported by a robot. Many robots now come with pipetting heads that can be exchanged to fit the customer's needs (small volume vs. large volume pipetting). Speed of the robot is an important consideration and a 96-probe pipetter can make a tremendous difference in set-up time when compared to an 8-channel tip. A crucial factor when considering buying any robotic equipment is to account for operation and maintenance cost of the machine. The software supplied with the robot is not only a critical determinant in day-to-day operation but it also has to be able to integrate with other robotic systems and other databases in the laboratory. The software packages distributed with the robotic equipment can range from easy-to-use software (usually less flexible) to software that might require actual code-writing (offering more flexibility to the advanced user).

3.6.2. Genotyping Instruments

Another rate-limiting step is the electrophoresis of PCR products to resolve and accurately size the different alleles. Fast electrophoresis runs and real-time detection of PCR products are possible using automated sequencers. Such fluorescence-based methods automate gel analysis by providing on-line signal detection as well as integrated software for fragment sizing. While this does not fully automate the allele-calling process, the software for these instruments provides for easy error checking and formatting of the data for input into linkage programs. Automation of the electrophoresis step will bring immediate time and personnel saving, produces the most accurate data for microsatellite typing and is cost-efficient since PCR products can be pooled for gel loading. Having integrated software that also provides on-line sample tracking minimizes the risk of errors during manual reading and transcription of the data and increases the throughput in data analysis. These instruments can be considered the most important step towards a semiautomated linkage analysis pipeline.

Currently, DNA gel electrophoresis hardware that incorporates software for the analysis of microsatellite markers is available from PE Applied Biosystems and Licor. Some groups have constructed their own instrumentation to accomplish high-throughput genotyping (J. Weber, unpubl.). Newer instruments that are based on capillary electrophoresis are described below (see Sect. 3.7.1). Table 3.3 compares the two most widely used instruments (ABI 377 and Licor

Table 3.3. Specifications of some commercially available genotyping instruments

	PE Applied Biosystems ABI 377	PE Applied Biosystems ABI 3700	Licor IR2	Molecular Dynamics MegaBase 1000
Capacity	96 Lanes	96 Lanes (384 upgrade available soon)	96 Lanes	96 Lanes
Typical run time	2–3 h	2–3 h (shorter capillary will be available in the future)	2–3 h	2–3 h
Base-pair resolution	1 bp	1 bp	1 bp	
Typical daily throughput (assume: 10 × multiplexing for ABI, 7 × multiplexing for Licor)	3 Runs/day (2880 genotypes)	6 Runs/day (5760 genotypes); expansion under development	4 Runs/day assuming reloading of gels (2240 genotypes)	4 Runs/day (3840 genotypes); planned robotic feeding of plates will increase genotyping capacity
Features	Visible fluorescent	Visible fluorescent	Infrared fluorescence	Visible fluorescent
Software	Genotyper Database extension	GeneScan Genotyper Database extension	SAGA or GeneImagIR Database extension	Expected release in 4Q'99
Summary	User-friendly software	User-friendly software	User-friendly software	Not tested

IR2) and lists some of the criteria that are important when considering buying this type of equipment for a microsatellite genotyping laboratory. Two additional instruments based on capillary electrophoresis that have only just been released for genotyping applications (ABI 3700 and Molecular Dynamics MegaBase1000) are listed as well. All the machines use the same principle for on-line data collection but detect the signals from the PCR amplification at different wavelengths. This is not such an important difference since PCR primers can easily be synthesized to accommodate both setups. However, at the moment there are many dyes available with emission peaks in the visible region of the spectrum, although not all of them can be used together in the same run since their emission spectra overlap.

The most important factor is certainly the throughput of the instrument (and also how much labor is involved in setting-up the instruments for a particular run). This is usually measured in genotypes per day and is determined by the number of gel lanes, the degree of multiplexing of PCR products (by size and primer color), the number of times the same gel can be reloaded, the electrophoresis run time, and the set-up time. It is difficult to make accurate estimates since many of these parameters will differ between laboratories. For example, the degree of multiplexing is determined by the population under study and the run times are determined by the required base-pair resolution. The throughput can always be increased by performing shorter runs on shorter gels. Another distinguishing factor will be the software provided with a particular instrument. The software is specifically designed to interpret the raw data of a particular instrument and supplied directly by the manufacturer. Other aspects of the software support the data analysis and offer more flexibility. Only very few laboratories will have the necessary bioinformatics support to rewrite portions of this software. It will therefore be important to include this factor when making a decision about purchasing a particular instrument and carefully evaluate the different user-interfaces. In addition, it should be noted that the instruments described here are very flexible and can also be used to support sequencing as well as mutation detection.

Commonly, fluorescent PCR products are generated using primers with the appropriate dyes attached. These signals are then detected by lasers that scan the gels during electrophoresis. Both the ABI 377 and the Licor IR2 described in Table 3.3 are equally suitable for large-scale genotyping setups. They support the same number of lanes per gel (96) but differ in the number of dyes that are currently available for a given emission spectrum. For both instruments, the number of genotypes per day is comparable. The software provided by the manufacturers differs markedly from a user's perspective but essentially accomplishes the same. Individual preference will probably be the most important factor in deciding which instrument to buy. Details of capillary instruments and their use for genotyping are described in Section 3.7.1.

3.7 Future Developments

Although significant improvements have been made to increase sample throughput for microsatellite analysis using gel-based systems, in order to achieve ultrahigh throughput, more revolutionary approaches are required. Various non-gel-based platform technologies are being developed for microsatellite analysis such as capillary-based systems and mass spectrometry based approaches. These developments, together with exciting improvements in nanotechnologies that may be applicable to microsatellite analysis, are discussed below.

3.7.1 Capillary Array Electrophoresis

At present, the application of capillary array electrophoresis (CAE) to microsatellite analysis is one of the most promising platform technologies for fulfilling the increasing demands in throughput. Capillary-based electrophoresis has several advantages over conventional gel electrophoresis. These include higher sensitivity (hence smaller sample size) and rapid run times as well as superior separation performance (Mansfield et al. 1996). In addition, various labor-intensive steps required in gel-based assays are eliminated in CAE. For example, gel preparation is substituted by automated capillary filling; manual sample loading of gels is replaced by an auto-injection mechanism; and the tedious gel lane tracking process is not required in CAE. The throughput of CAE is dependent on the number of capillaries in the array, and whether the electrophoresis process is automated. Using a typical capillary electrophoresis machine, it takes about 2–3 h to perform a run (including sample preparation time). The CEQ2000 (Beckman Coulter, Fullerton) has an array of eight capillaries, and performs automated thermal denaturation prior to automatic injection. This machine has an estimated throughput of 640 genotypes per day (assuming multiplexing of 10 markers) with 8 unattended runs per day. The MegaBACE 1000 (Molecular Dynamics Inc, Sunnyvale) is an example of a 96-capillary sequencer that processes 96 samples automatically. A CAE machine with 96 capillaries can generate up to 3840 genotypes per day, with 4 runs per working day. In addition, CAE machines such as the ABI3700 (PE Applied Biosystems) with 104 capillaries (96 working and 8 reserve capillaries) have additional plate storage capacity (up to four 96- or 384-well plates) and automated loading will allow 24 h unattended operation, generating up to 7,680 genotypes per machine per day. Other CAE machines that are currently under development include microfabricated 96-sample capillary array electrophoresis microplates (DNA Sciences, Mountain View; Simpson et al. 1998). It can be anticipated that a significant reduction in cost will be achieved using CAE machines due to the decrease in sample volume required for analysis (see discussion in Sect. 3.7.3). In addition, considerable savings in labor costs for the various preparation steps in slab-gel electrophoresis will also decrease the cost of genotyping. With the development of shorter capillaries, and perhaps even higher numbers of capillaries in the array, the genotyping capacity of the CAE machines will approach ultrahigh throughput (Scherer et al. 1999). The increased capacity, coupled with the reduction in cost per genotype, will facilitate large-scale genetic studies in the near future.

3.7.2 Mass Spectrometry

An alternative non-gel-based approach for fragment size analysis is to determine the molecular weight of a given nucleic acid using mass spectrometry. The attraction of mass spectrometry based machines is the precision and accuracy of mass measurements. Furthermore, as the instrument is very sensitive, the amount of sample required for analysis is greatly reduced, thus reducing the cost of each genotype. In addition, the running time of mass spectrometry is <1 s, thus providing the higher throughput required for many applications. Matrix-assisted laser desorption/ionization time-of-flight mass spectrometry (MALDI-TOFMS), as well as electrospray ionization (ESI) mass spectrometry, have been used to perform accurate DNA fragment size analysis (Ross and Belgrader 1997; Wada 1997; Laken et al. 1998). Furthermore, microsatellite typing using TOFMS has been demonstrated using systems developed by companies such as Sequenom (Braun et al. 1997; Tang et al. 1999) as well as Gene-Trace Systems (Monforte and Becker 1997; Butler et al. 1999).

In addition, adaptations such as using MassTag-labeled primers (PerSeptive Biosystems, Framingham) or cleavable mass spectrometry tags (CMSTs; Rapigene Inc., Bothell) allow the multiplexing of assays. MassTag-labeled primers are primers with 5'-oligo(dT) sequences that allow discrimination of peaks by mass in a mass spectrometry run containing multiple DNA fragments (Haff and Smirnov 1997). A similar strategy for multiplexing is employed using CMSTs. These tags are small molecules attached by a linker to the 5' end of the primer. Upon exposure to light, the linker is cleaved away and the tag with the known molecular weight can be detected by mass spectrometry (Steinberg 1998). At present, both approaches to multiplexing are via mass discrimination provided by the mass tags. By running several mass tags simultaneously in a single mass spectrometry 'run', parallel processing of many genotyping reactions can be achieved. These approaches are at present mainly applied to SNP analysis, but may also be adaptable to microsatellite typing when improvements in increasing the length of DNA to be analyzed by mass spectrometry are finally achieved (see below).

At present, there is a limit to the size of DNA that can be analyzed (<100 bases), due to the random fragmentation of DNA during mass spectrometry. This poses a major problem for microsatellite analysis as most DNA fragments to be analyzed are greater than 100 bp. Various implementations, such as incorporating 7-deaza nucleotides (Schneider and Chait 1995) or using infrared

MALDI (Berkenkamp et al. 1998), will allow the analysis of DNA fragments up to 2000 bp. Another challenge for MALDI-TOF analysis is the reduction of the sample preparation prior to the mass spectrometry step. Although the measurement of molecular weight by mass spectrometry takes only a few seconds per sample, post-PCR sample preparation is laborious and time consuming. Typically, for MALDI-TOFMS analysis, the DNA fragments have to be purified, desalted and concentrated prior to suspension in the matrix solution ready for mass spectrometry. For ESI-MS, DNA samples have to undergo a similar clean-up procedure prior to loading onto a high performance liquid chromatograph coupled to the mass spectrometer. Automation of the sample preparation stage is under development in companies such as Sequenom (San Diego) and GeneTrace Systems (Menlo Park). Although improvements can be made to allow large-scale microsatellite typing using mass spectrometry, it appears that mass spectrometry approaches are much better suited for SNP identification and SNP typing.

3.7.3 Nanotechnologies

The main driving force behind scaling down PCR reactions towards the nanoscale is to reduce cost. Furthermore, as there is usually a limited amount of DNA sample available for analysis, nanoscale reactions will enable more genotypes per DNA sample to be performed and allow for a reduction in sample consumption. In addition, the ability to parallel process a 100-fold more DNA samples will not only make the genotyping process more cost effective but will also lead to further increases in throughput.

The move towards nanoscale reactions poses new challenges. DNA in solution is difficult to handle at the nanoscale as the DNA molecules are very rigid; evaporation will be significant at microscales and special surfaces will be required to ensure that the liquid does not 'stick' to the reaction vessel due to surface tension or as a result of 'bridging' of liquids between wells due to capillary action. Furthermore, in order to perform nanoscale fluid manipulations, special liquid handling systems are required to counteract the effect of surface tension for accurate liquid transfers of very small volumes. For example, the Nano-Plotter (GeSiM mbH) and nQUAD nanolitre liquid handler (Cartesian Technologies) are micropipetting systems developed to handle samples in the nanoliter range. The Nano-Plotter has piezoelectric tips for precise dispensing of small volumes of fluids, whereas the nQUAD system utilizes a high-speed microsolenoid valve to dispense fluids via changes in pressure and speed.

Another approach toward miniaturization is the integration of all of the DNA assays into a single platform. This allows the handling of all fluids in an enclosed environment, hence eliminating the effect of evaporation. In addition, single systems enable full automation of the genotyping process and also facilitate the transfer of fluids from one machine to the next. For example, an integrated system for genotyping directly from blood has been developed at the

Ames laboratory (Zhang et al. 1999). The blood sample is loaded into a fused-silica capillary, and PCR is performed using a hot-air thermal cycler. The products are automatically loaded onto the attached capillary electrophoresis machine for fragment sizing. Another example of an integrated system is based on the Caliper LabChip technology (commercial partner: Agilent Technologies). The first generation LabChip was designed to perform automated DNA size analysis on a single microfluidic chip. This system has integrated sample handling, separation and detection within a single platform. Integrated systems for microsatellite analysis will facilitate DNA typing for various applications such as molecular diagnostics and in forensic medicine. Although all of these systems are still at the pilot stage of development, miniaturization and integrated systems will become increasingly important in the near future (Service 1998).

3.8 Applications

The availability of large numbers of microsatellite markers, and their highly informative nature, have provided us with a great resource for human genetic analysis. Advancements in genotyping technologies have accelerated the speed of mapping the human genome and greatly facilitated the isolation of disease genes by positional cloning. At present, the comprehensive human genetic map, which comprises about 8000 highly informative microsatellite markers, allows the precise localization of a gene within a 1–2 Mb interval (Broman et al. 1998; Weissenbach 1998). Linkage analysis using microsatellite markers has proven to be successful in the isolation of genes for monogenic diseases such as cystic fibrosis (Riordan et al. 1989) and Huntington's chorea (Huntington's Disease Collaborative Research Group 1993). Genome-wide scans performed using evenly spaced microsatellite markers yielded initial success in demonstrating linkage to genetic loci for a variety of complex diseases such as multiple sclerosis (Sawcer et al. 1996) and obesity (Hager et al. 1998). Once a disease locus has been narrowed down to a 'minimal critical region' by linkage analysis, the gene can be localized by fine mapping.

One approach for finer localization of a disease gene is to use linkage disequilibrium (LD), otherwise known as allelic association approaches. In LD studies, information from meiotic events throughout many generations is evaluated under the assumption that the disease mutation occurred in one individual as a founder effect. This approach is particularly suited for disease gene mapping using isolated populations. Due to the highly polymorphic nature of microsatellite markers, they have been widely used in the identification of several rare recessive disease loci and genes in population isolates, for example, the isolation of the Hirschsprung disease (HSCR) susceptibility locus (HSCR2) in Old Order Mennonites of Lancaster County, Pennsylvania (Puffenberger et al. 1994) and the identification of the gene responsible for progressive myoclonus epilepsy, based on LD analysis using the Finnish population (Pennacchio et al. 1996). In addition, microsatellites are used in deletion mapping studies based on loss of heterozygosity (LOH) in tumors. LOH or allelic imbalance represents the addition or loss of genetic material in tumor samples, thus providing circumstantial evidence for the location of cancer-related genes. Various cancer-susceptibility loci have been identified using the LOH approach, leading to the identification of various tumor-suppressor genes such as PTEN (Li et al. 1997) and MEN1 (Chandrasekharappa et al. 1997).

Microsatellite markers are also widely used in human population genetics such as the study of phylogenetic relationships among closely related individuals or populations and human evolutionary history (Bowcock et al. 1994; Cavalli-Sforza 1998; Chakravarti 1999). In addition, microsatellites can also be used for DNA fingerprinting, and for various forensics applications (see Chap. 4). Finally, application of microsatellite markers is not only restricted to applications in human genetics. These markers are also widely used in genetic studies of various model organisms. The development of high-resolution genetic linkage maps using microsatellite markers for various model organisms such as mouse and rat (McCarthy et al. 1997; Steen et al. 1999) provides exciting prospects for understanding mammalian genome evolution through comparative mapping, for developing animal models of human diseases, for developmental and physiological pathway analysis and, ultimately, towards the understanding of gene function.

References

Beckman JS, Weber JL (1992) Survey of human and rat microsatellites. Genomics 12:627–631

Berkenkamp S, Kirpekar F, Hillenkamp F (1998) Infrared MALDI mass spectrometry of large nucleic acids. Science 281:260–262

Botstein D, White RL, Skolnick M, Davis RW (1980) Construction of a genetic linkage map in man using restriction fragment lengths polymorphisms. Am J Hum Genet 32:314–331

Bowcock AM, Ruiz-Linares A, Tomfohrde J, Minch E, Kidd JR, Cavalli-Sforza LL (1994) High resolution of human evolutionary trees with polymorphic microsatellites. Nature 368:455–457

Braun A, Little DP, Reuter D, Müller-Mysok B, Köster H (1997) Improved analysis of microsatellites using mass spectrometry. Genomics 46:18–23

Broman KW, Murray JC, Sheffield VC, White RL, Weber JL (1998) Comprehensive human genetic maps: individual and sex-specific variation in recombination. Am J Hum Genet 63:861–869

Brownstein MJ, Carpten JD, Smith JR (1996) Modulation of non-templated nucleotide addition by Taq DNA polymerase: primer modifications that facilitate genotyping. Biotechniques 20: 1004–1010

Brzustowicz LM, Merette C, Xie X, Townsend L, Gilliam TC, Ott J (1993) Molecular and statistical approaches to the detection and correction of errors in genotype databases. Am J Hum Genet 53:1137–1145

Buetow KH (1991) Influence of aberrant observations on high-resolution linkage analysis outcomes. Am J Hum Genet 49:985–994

Buetow KH, Weber JL, Ludwigsen S, Scherpbier-Heddema T, Duyk GM, Sheffield VC, Wang Z, Murray JC (1994) Integrated human genome-wide maps constructed using the CEPH reference panel. Nat Genet 6:391–393

Butler JM, Li J, Shaler TA, Monforte JA, Becker CH (1999) Reliable genotyping of short tandem repeat loci without an allelic ladder using time-of-flight mass spectrometry. Int J Legal Med 112:45–49

Callen DF, Thompson AD, Shenm Y, Phillips HA, Richards RI, Mulley JC, Sutherland GR (1993) Incidence and origin of "null" alleles in the (AC)n microsatellite markers. Am J Hum Genet 52:922–927

Cassel SM, Guttman A (1998) Membrane-mediated sample loading for automated DNA sequencing. Electrophoresis 19:1341–1346

Cavalli-Sforza LL (1998) The DNA revolution in population genetics. Trends Genet 14:60–65

Chakravarti A (1999) Population genetics – making sense out of sequence. Nat Genet 21 [Suppl 1]:56–60

Chandrasekharappa SC, Guru SC, Manickam P, Olufemi S-E, Collins FS, Emmert-Buck MR, Debelenko LV, Zhuang Z, Lubensky IA, Liotta LA, Crabtree JS, Wang Y, Roe BA, Weisemann J, Boguski MS, Agarwal SK, Kester MB, Kim YS, Heppner C, Dong Q, Spiegel AM, Burns AL, Marx SJ. (1997) Positional cloning of the gene for multiple endocrine neoplasia-type 1. Science 276: 404–406

Cooper ML, Maffitt DR, Parsons JD, Hillier L, States DJ (1996) Lane tracking software for four-colour fluorescence-based electrophoretic gel images. Genome Res 6:1110–1117

Cottingham RW Jr, Idury RM, Schaffer AA (1993) Faster sequential genetic linkage computations. Am J Hum Genet 53:252–263

Dib C, Faure S, Fizames C, Samson D, Drouot N, Vignal A, Millasseau P, Marc S, Hazan J, Seboun E, Lathrop M, Gyapay G, Morissette J, Weissenbach J (1996) A comprehensive genetic map of the human genome based on 5264 microsatellites. Nature 380:152–154

Don RH, Cox PT, Wainwright BJ, Baker K, Mattick JS (1991) 'Touchdown' PCR to circumvent spurious priming during gene amplification. Nucleic Acids Res 19:4008

Donis-Keller H, Green P, Helms C, Cartinhour S, Weiffenbach B, Stephens K, Keith TP, Bowden DW, Smith DR, Lander ES (1987) A genetic linkage map of the human genome. Cell 51:319–337

Dracopoli NC, Connell PO, Elsner TI, Lalouel J, White RL, Buetow KH, Nishimura DY, Murray JC, Helms C, Mishra SK, Donis-Keller H, Hall JM, Lee MK, King MC, Attwood J, Morton NE, Robson EB, Mahtani M, Willard HF, Royle NJ, Patel I, Jeffrey AJ, Verga V, Jenkins T, Weber JL, Mitchell AL, Bale AE (1991) The CEPH consortium linkage map of human chromosome 1. Genomics 9:686–700

Ehm MG, Kimmel M, Cottingham RW Jr (1995) Error detection in genetic linkage data for human pedigrees using likelihood ratio methods. J Biol Syst 3:13–25

Ehm MG, Kimmel M, Cottingham RW Jr (1996) Error detection for genetic data, using likelihood methods. Am J Hum Genet 58:225–234

Erfle H, Ventzki R, Voss H, Rechmann S, Benes V, Stegemann J, Ansorge W (1997) Simultaneous loading of 200 sample lanes for DNA sequencing on vertical and horizontal, standard and ultra thin gels. Nucleic Acids Res 25:2229–2230

Ghosh S, Karanjawala ZE, Hauser ER, Ally DS, Knapp JI, Rayman JB, Musick A, Tannenbaum J, Te C, Shapiro S, Eldridge W, Musick T, Martin C, Smith JR, Carpten JD, Brownstein MJ, Powell JI, Whiten R, Chines P, Nylund SJ, Magnuson VL, Boehnke M, Collins FS (1997) Methods for precise sizing, automated binning of alleles, and reduction of error rates in large-scale genotyping using fluoroscently labeled dinucleotide markers. FUSION (Finland-US Investigation of NIDDM Genetics) Study Group. Genome Res 7:165–178

Ghosh S, Watanabe RM, Hauser ER, Valle T, Magnuson VL, Erdos MR, Langefeld CD, Balow J Jr, Ally DS, Kohtamaki K, Chines P, Birznieks G, Kaleta H-S, Musick A, Te C, Tannenbaum J, Eldridge W, Shapiro S, Martin C, Witt A, So A, Chang J, Shurtleff B, Porter R, Kudelko K, Unni A, Segal L, Sharaf R, Blaschak-Harvan J, Eriksson J, Tenkula T, Vidgren G, Ehnholm C, Tuomilehto-Wolf E, Hagopian W, Buchanan TA, Tuomilehto J, Bergman RN, Collins FS, Boehnke M (1999) Type 2 diabetes: evidence for linkage on chromosome 20 in 716 Finnish affected sib pairs. Proc Natl Acad Sci USA 96:2198–2203

Ginot F, Bordelais I, Nguyuen S, Gyapay G (1996) Correction of some genotyping errors in automated fluorescent microsatellite analysis by enzymatic removal of one base overhangs. Nucleic Acids Res 24:540–541

Golden JB, Togersen D, Tibbetts C (1993) On-line signal conditioning and feature extraction for basecalling. Intelligent Sys Mol Biol 1:136–144

Goldstein DR, Zhao H, Speed TP (1997) The effects of genotyping errors and interference on estimation of genetic distance. Hum Hered 47:86–100

Green P, Falls K, Crooks S (1988) Documentation for CRI-MAP, version 2.1. Department of Human Genetics, Collaborative Research Inc., Bedford, MA 01730, USA

Grimalidi MC, Crouau-Roy B (1997) Microsatellite allelic homoplasy due to variable flanking sequences. J Mol Evol 44:336–340

Guttman A (1999) Sample stacking during membrane-mediated loading in automated DNA sequencing. Anal Chem 71:3598–3602

Gyapay G, Morissette J, Vignal A, Dib C, Fizames C, Millasseau P, Marc S, Bernardi G, Lathrop M, Weissenbach J (1994) The 1993–94 Genethon human genetic linkage map. Nat Genet 7: 246–339

Haff LA, Smirnov IP (1997) Multiplex genotyping of PCR products with MassTag-labeled primers. Nucleic Acids Res 25:3749–3750

Hager J, Dina C, Francke S, Dubois S, Houari M, Vatin V, Vaillant E, Lorentz N, Basdevant A, Clement K, Guy-Grand B, Froguel P (1998) A genome-wide scan for human obesity genes reveals a major susceptibility locus on chromosome 10. Nat Genet 20:304–308

Hall JM, LeDuc CA, Watson AR, Roter AH (1996) An approach to high-throughput genotyping. Genome Res 6:781–790

Hauge XY, Litt M (1993) A study of the origin of 'shadow bands' seen when typing dinucleotide repeat polymorphisms by the PCR. Hum Mol Genet 2:411–415

Huntington's Disease Collaborative Research Group (1993) A novel gene containing a trinucleotide repeat that is expanded and unstable on Huntington's disease chromosomes. Cell 72:971–983

Idury RM, Cardon LR (1997) A simple method for automated allele binning in microsatellite markers. Genome Res 7:1104–1109

Koorey DJ, Bishop GA, McCaughan GW (1993) Allele non-amplification; a source of confusion in linkage studies employing microsatellite polymorphisms. Hum Mol Genet 2:289–291

Kruglyak L, Daly MJ, Reeve-Daly MP, Lander ES (1996) Parametric and nonparametric linkage analysis : a unified multipoint approach. Am J Hum Genet 58:1347–1363

Kwok S, Higuchi R (1989) Avoiding false positives with PCR. Nature 339:237–238

Laken SJ, Jackson PE, Kinzler KW, Vogelstein B, Strickland PT, Groopman JD, Friesen MD (1998) Genotyping by mass spectrometric analysis of short DNA fragment. Nat Biotechnol 16: 1352–1356

Lander ES, Green P (1987) Construction of multilocus linkage maps in humans. Proc Natl Acad Sci USA 84:2363–2367

Lange K, Weeks D, Boehnke M (1998) Programs for pedigree analysis: MENDEL, FISHER, and dGENE. Genet Epidemiol 5:471–472

Lasher L, Reefer J, Chakravarti A (1991) Effects of genotyping errors on the estimation of chromosome map length. Am J Hum Genet Suppl 49:A369

Lathrop GM, Huntsman JW, Hooper AB, Ward RH (1983) Evaluating pedigree data: identifying the cause of error in families with inconsistencies. Hum Hered 33:377–389

Levitt RC, Kiser MB, Dragwa C, Jedlicka AE, Xu J, Meyers DA, Hudson JR (1994) Fluorescence-based resource for semiautomated genomic analyses using microsatellite markers. Genomics 24:361–365

Li J, Yen C, Liaw D, Podsypanina K, Bose S, Wang SI, Puc J, Miliaresis C, Rodgers L, McCombie R, Bigner SH, Giovanella BC, Ittmann M, Tycko B, Hibshoosh H, Wigler MH, Parsons R (1997) PTEN, a putative protein tyrosine phosphatase gene mutated in human brain, breast, and prostate cancer. Science 275:1943–1947

Lin Z, Xiangfeng C, Li H (1996) Multiplex genotype determination at a large number of gene loci. Proc Natl Acad Sci USA 93:2582–2587

Lincoln SE, Lander ES (1992) Systematic detection of errors in genetic linkage data. Genomics 14:604–610

Lindqvist AK, Magnusson PK, Balciuniene J, Wadelius C, Lindholm E, Alarcon-Riquelme ME, Gyllensten UB (1996) Chromosome-specific panels of tri- and tetranucleotide microsatellite markers for multiplex fluorescent detection and automated genotyping: evaluation of their utility in pathology and forensics. Genome Res 6:1170–1176

Litt M, Luty JA (1989) A hypervariable microsatellite revealed by in vitro amplification of a dinucleotide repeat within the cardiac muscle actin gene. Am J Hum Genet 44:397–401

Magnuson VL, Ally DS, Nylund SJ, Karanjawala ZE, Rayman JB, Knapp JI, Lowe AL, Ghosh S, Collins FS (1996) Substrate nucleotide-determined non-templated addition of adenine by Taq DNA polymerase: implications for PCR-based genotyping and cloning. Biotechniques 21:700–709

Mansfield ES, Vainer M, Enad S, Barker DL, Harris D, Rappaport E, Fortina P (1996) Genome Res 6:893–903

Matise RC, Perlin M, Chakravarti A (1994) Automated construction of genetic linkage maps using an expert system (MultiMap): a human genome linkage map. Nat Genet 6:384–390

Mayrand PE, Corcoran KP, Ziegle JS, Roberson JM, Hoff LB, Kronick MN (1992) The use of fluorescence detection and internal lane standards to size PCR products automatically. Appl Theor Electrophor 3:1–11

McCarthy LC, Terrett J, Davies ME, Knights CJ, Smith AL, Critcher R, Schmitt K, Hudson J, Spurr NK, Goodfellow PN (1997) A first-generation whole genome radiation hybrid map spanning the mouse genome. Genome Res 7:1153–1161

Miller MJ, Yuan BZ (1997) Semi automated resolution of overlapping stutter patterns in genomic microsatellite analysis. Anal Biochem 251:50–56

Monforte JA, Becker CH (1997) High-throughput DNA analysis by time-of-flight mass spectrometry. Nat Med 3:360–362

Pennacchio LA, Lehesjoki AE, Stone NE, Willour VL, Virtaneva K, Miao J, D'Amato E, Ramirez L, Faham M, Koskiniemi M, Warrington JA, Norio R, de la Chapelle A, Cox DR, Myers RM (1996) Mutations in the gene encoding cystatin B in progressive myoclonus epilepsy. Science 271:1731–1734

Perlin MW, Lancia G, Ng SK (1995) Toward fully automated genotyping: genotyping microsatellite markers by deconvolution. Am J Hum Genet 57:1199–1210

Puffenberger EG, Hosoda K, Washinton SS, Nakao K, deWit D, Yanagisawa M, Chakravarti A (1994) A missense mutation on the endothelin-B receptor gene in multigenic Hirschsprung's disease. Cell 79:1279–1266

Reed PW, Davies JL, Copeman JB, Bennett ST, Palmer SM, Pritchard LE, Gough SCL, Kawaguchi Y, Cordell HJ, Balfour KM, Jenkins SC, Powell EE, Vignal A, Todd JA (1994) Chromosome-specific microsatellite sets for fluorescence-based, semi-automated genome mapping. Nat Genet 7:390–395

Riordan JR, Rommens JM, Kerem B, Alon N, Rozmahel R, Grzelczak Z, Zielenski J, Lok S, Plavsic N, Chou JL, Drumm ML, Iannuzzi MC, Collins FS, Tsui L-C (1989) Identification of the cystic fibrosis gene: cloning and characterization of complementary DNA. Science 245:1066–1073

Risch N, Spiker D, Lotspeich L, Nouri N, Hinds D, Hallmayer J, Kalaydjieva L, McCague P, Dimiceli S, Pitts T, Nguyen L, Yang J, Harper C, Thorpe D, Vermeer S, Young H, Hebert J, Lin A, Ferguson J, Chiotti C, Wiese-Slater S, Rogers T, Salmon B, Nicholas P, Myers RM (1999) A genomic screen of autism: evidence for a multilocus etiology. Am J Hum Genet 65:493–507

Ross PL, Belgrader P (1997) Analysis of short tandem repeat polymorphisms in human DNA by matrix-assisted laser desorption/ionization mass spectrometry. Anal Chem 69:3966–3972

Sawcer S, Jones HB, Feakes R, Gray J, Smaldon N, Chataway J, Robertson N, Clayton D, Goodfellow PN, Compston A (1996) A genome screen in multiple-sclerosis reveals susceptibility loci on chromosome 6p21 and 17q22. Nat Genet 13:464–468

Scherer JR, Kheterpal I, Radhakrishnan A, Ja WW, Mathies RA (1999) Ultra-high throughput rotary capillary array electrophoresis scanner for fluorescent DNA sequencing and analysis. Electrophoresis 20:1508–1017

Schneider K, Chait BT (1995) Increased stability of nucleic acids containing 7-deaza-guanosine and 7-deaza-adenosine may enable rapid DNA sequencing by matrix-assisted laser desorption mass spectrometry. Nucleic Acids Res 23:1570–1575

Service RF (1998) Coming soon: the pocket DNA sequencer. Science 282:399–401

Shadle SE, Allen DF, Guo H, Pogozelski WK, Bashkin JS, Tullius TD (1997) Quantitative analysis of electrophoresis data: novel curve fitting methodology and its application to the determination of a protein-DNA binding constant. Nucleic Acids Res 25:850–860

Sheffield VC, Weber JL, Buetow KH, Murray JC, Even DA, Wiles K, Gastier JM, Pulido JC, Yandava C, Sunden SL, Mattes G, Businga T, McClain A, Beck J, Scherpier T, Gilliam J, Zhong J, Duyk GM (1995) A collection of tri- and tetranucleotide repeat markers used to generate high quality, high resolution human genome-wide linkage maps. Hum Mol Genet 4:1837–1844

Shield DC, Collins A, Buetow KJ, Morton NE (1991) Error filtration, interference, and the human linkage map. Proc Natl Acad Sci USA 88:6501–6505

Shuber F, Grondin V, Klinger K (1995) A simplified procedure for developing multiplex PCRs. Genome Res 5:488–493

Simpson PC, Roach D, Woolley AT, Thorsen T, Johnston R, Sensabaugh GF, Mathies RA (1998) High-throughput genetic analysis using microfabricated 96-sample capillary array electrophoresis microplates. Proc Natl Acad Sci USA 95:2256–2261

Smith JR, Carpten JD, Brownstein MJ, Ghosh S, Magnuson VL, Gilbert DA, Trent JM, Collins FS (1995) Approach to genotyping errors caused by nontemplated nucleotide addition by Taq DNA polymerase. Genome Res 5:312–317

Stallings RL, Ford AF, Nelson D, Torney DC, Hildebrand CE, Moyzis RK (1991) Evolution and distribution of (GT)n repetitive DNA sequences in mammalian genomes. Genomics 10: 807–815

Steen RG, Kwitek-Black AE, Glenn C, Gullings-Handley J, Van Etten W, Atkinson OS, Appel D, Twigger S, Muir M, Mull T, Granados M, Kissebah M, Russo K, Crane R, Popp M, Peden M, Matise T, Brown DM, Lu J, Kingsmore S, Tonellato PJ, Rozen S, Slonim D, Young P, Knoblauch M, Provoost A, Ganten D, Colman SD, Rothberg J, Lander ES, Jacob HJ (1999) A high-density genetic integrated linkage and radiation hybrid map of the laboratory rat. Genome Res 9: AP1–AP8

Steinberg D (1998) Hybridisation buffers and novel DNA tags developed. Genetic Engineering News 18

Stringham HM, Boehnke M (1996) Identifying marker typing incompatibilities in linkage analysis. Am J Hum Genet 59:946–950

Tang K, Fu DJ, Julien D, Braun A, Cantor CR, Koster H (1999) Chip-based genotyping by mass spectrometry. Proc Natl Acad Sci USA 96:10016–10020

Tautz D (1989) Hypervariability of simple sequences as a general source for polymorphic DNA markers. Nucleic Acids Res 17:6463–6471

Telenius H, Carter NP, Bebb CE, Nordenskjold M, Ponder BA, Tunnacliffe A (1992) Degenerate oligonucleotide-primed PCR: general amplification of target DNA by a single degenerate primer. Genomics 13:718–725

Wada Y (1997) Separate analysis of complementary strands of restriction enzyme-digested DNA. An application of restriction fragment mass mapping by matrix-assisted laser desorption/ionization mass spectrometry. J Mass Spectrom 33:197–192

Wang DG, Fan JB, Siao CJ, Berno A, Young P, Sapolsky R, Ghandour G, Perkins N, Winchester E, Spencer J, Kruglyak L, Stein L, Hsie L, Topaloglou T, Hubbell E, Robinson E, Mittmann M, Morris MS, Shen N, Kilburn D, Rioux J, Nusbaum C, Rozen S, Hudson TJ, Lander ES (1998) Large-scale identification, mapping, and genotyping of single-nucleotide polymorphisms in the human genome. Science 280:1077–1082

Weber JL, May PE (1989) Abundant class of human DNA polymorphisms which can be typed using the polymerase chain reaction. Am J Hum Genet 44:388–396

Weber JL, Wong C (1993) Mutation of human short tandem repeats. Hum Mol Genet 2:1123–1128

Weissenbach J (1998) The human genome project : from mapping to sequencing. Clin Chem Lab Med 36:511–514

Weissenbach J, Gyapay G, Dib C, Vignal A, Morissette J, Millasseau P, Vaysseix G, Lathrop M (1992) A second-generation linkage map of the human genome. Nature 359:794–801

Xu X, Rogus JJ, Terwedow HA, Yang J, Wang Z, Chen C, Niu T, Wang B, Xu H, Weiss S, Schork NJ, Fang Z (1999) An extreme-sib-pair genome scan for gene regulating blood pressure. Am J Hum Genet 64:1694–1701

Yuan B, Vaske D, Weber JL, Beck J, Sheffield VC (1997) Improved set of short-tandem-repeat polymorphisms for screening the human genome. Am J Hum Genet 60:459–460

Zhang L, Cui X, Schmitt K, Hubert R, Navidi W, Arnheim N (1992) Whole genome amplification from a single cell: implications for genetic analysis. Proc Natl Acad Sci USA 89:5847–5851

Zhang N, Tan H, Yeung ES (1999) Automated and integrated system for high-throughput DNA genotyping directly from blood. Anal Chem 71:1138–1145

4 Minisatellite and Microsatellite DNA Fingerprinting

PAUL G. DEBENHAM

4.1 Introduction

Most DNA-based inventions receive much media attention, but have little to show in the end by way of practical outcome. One major exception is DNA fingerprinting and the subsequent variations in mini/microsatellite applications. The discovery of minisatellite DNA fingerprinting by A. Jeffreys in 1985 (Jeffreys et al. 1985a,b) in one brief step revolutionised the way forensic science and police casework is performed, and became the norm for the determination of relationships for both humans and animals.

The technology and applications have continued to evolve since 1985 and the terminology of DNA fingerprinting, multilocus probes, DNA profiling, single locus probes, variable nucleotide tandem repeats (VNTRs), minisatellites, microsatellites, and, more recently, short tandem repeats has become confusing and misused. In fact, there has been a logical progression in methodology through the years, although, as will be shown, the uptake and application of improvements have occasionally had a political edge to them.

4.2 Blood Grouping

Since the start of the 20th century, blood grouping has held a premier place as the forensic means of identification. By the time DNA fingerprinting was discovered, there were at least 17 different blood-typing systems (Stedman 1983) in use, some of which could be used for identification of non-blood tissues such as saliva or sperm. Blood groups were also the basis of paternity determinations, and order was brought to an ad hoc plethora of experts by the establishment in 1969 of Home Office Registered Paternity testers for the authorisation of blood grouping reports to the UK courts.

Despite its established status, blood grouping was not ideally suited to either forensic or paternity applications. While an array of different blood group markers could be highly discriminating, it was necessary to have a sufficiently large bloodstain to work with. Furthermore, environmental aging

Principles and Practice
Molecular Genetic Epidemiology – A Laboratory Perspective
Ian N.M. Day (Ed.)
© Springer-Verlag Berlin Heidelberg 2002

or microbial contamination rapidly depleted the antigens available for testing. Thus old or degraded samples often resulted in unsuccessful blood typing attempts.

Equally, whilst the use of an array of blood group types was usually effective at eliminating an unrelated man from being the father of a child, it was relatively ineffective in establishing the converse – that a man not excluded by blood grouping studies was necessarily the father of the child. This was because the range of types found for any particular blood group test was usually very limited and thus gave little positive proof of paternity. This was particularly of concern if related men were involved in disputed parentage.

The increasing sophistication of immunological methods and gel-electrophoretic separation chemistries were expanding the utility and repertoire of blood grouping methods. However, at the same time, research was equally improving the understanding of epitope detection and masking, post-translational modification of proteins, as well as carbohydrate biochemistry, such that the validity for identification purposes of such tests was increasingly questioned.

It was not surprising therefore that protein moieties would not be the basis of the ultimate identity test, especially as everyone knew this truly resided in DNA. However, the advent of identification via DNA analysis had to await a methodology that could overcome the technological hurdle that a human identity resided in ~3 billion DNA base pairs.

4.3 DNA Fingerprinting – The Discovery

Given the magnitude of the human genome and that the mean heterozygosity of DNA is low (approximately 0.001 per bp), the discovery of informative variation, without prior knowledge of the DNA sequence, still awaits the assessment and utility of the latest single nucleotide polymorphism methodologies in the 1990s. In 1980, however, a DNA sequence was discovered by chance that showed hypervariability between individuals and thus could act as a highly informative marker for the surrounding DNA region (Wyman and White 1980). Initially, this hypervariability was thought to be akin to that previously found in *E. coli* and yeast and derived from transposable elements. Subsequently, a few other hypervariable loci were uncovered and in each case the variable region contained side-by-side tandem repeats of a short sequence in the DNA. The hypervariability was associated with the number of repeats of the short sequence, which could differ between the two copies present in the individual, let alone between those in other people. The term minisatellite region was used to reflect the apparent high GC content of these sequences, which, through repetition, could distort density gradient distribution of fragmented DNA. The term variable nucleotide tandem repeat (VNTR) was also coined to refer to these sequences (Nakamura et al. 1987).

In 1984 A. Jeffreys identified a minisatellite region within the human myoglobin gene sequences he was studying. The 33-base-pair (bp) sequence involved was isolated and used as a probe with the intention of identifying the variation associated with the myoglobin gene structure in humans and other animals. In fact, what Jeffreys discovered under his experimental conditions was not the two-band pattern expected, reflecting the DNA repeat structure of this region in the two copies of an individual's DNA. Jeffreys discovered that his myoglobin minisatellite probe had numerous homologous matches in the DNA of individuals such that the resulting probe test illuminated a multitude of DNA bands in the DNA of the individual. Each band represented a hyper-variable region of similar minisatellite sequence (Jeffreys et al. 1985a). Jeffreys then made a further leap when he recognised that the resulting barcode-like pattern of bands, achieved by using his probe on human hypervariable DNA, was a graphic representation of human variability at the DNA level. The barcode provided a sufficiently detailed pattern for it to be used as a means of distinguishing between the DNA of individuals; a DNA (minisatellite) finger-print (Jeffreys et al. 1985b).

Jeffreys isolated a series of cloned sequences that hybridised with the myo-globin probe under low stringency conditions and from these identified two called 33.6 and 33.15. These seemed to act as probes for detecting minimally overlapping "families" of hypervariable loci (Jeffreys et al. 1986). Extensive studies have confirmed that, in the size range of DNA fragments analysed, only about 1% of the bands obtained using these probes represents the same loci (Jeffreys et al. 1991a). Essentially, 33.6 and 33.15 probes, the first of a breed of "multilocus" probes, gave independent DNA fingerprints which became the bedrock of DNA identification for the next few years.

The minisatellites hybridising to the probes had repeating sequences that ranged from approximately 9–40bp and existed in a multitude of repeated states such that the minisatellite could reach a size of 20kb. Below about 3.5kb, the density of fingerprint bands became too great to be reliably determined. However, between 4 and 20kb there existed in the order of 17 bands for either 33.6 or 33.15 to provide sufficient information to identify an individual (Jeffreys et al. 1991a). The intensity of the DNA bands after probing was also characteristic of the individual as it reflected the degree of homology with the probe. Therefore, the ability to accurately and reproducibly resolve the DNA fingerprint patterns became essential for a successful analysis of a DNA fingerprint.

The methodology involved in generating a DNA fingerprint relies on stan-dard molecular biological methodology, yet in practice few laboratories were able to master the technology necessary to give the reproducibility and defin-ition required for a forensic identification test. The test required ~2µg of high molecular weight DNA to be isolated from the sample usually using detergent lysis of the cells followed by proteinase K digestion of the non-nucleotide cellular structures. The DNA was removed from the resulting cell debris by phenol/chloroform partition of the aqueous phase followed by ethanol precipitation. The DNA was then spooled, dried, then resolvated in buffer suitable for restriction digestion. The choice of restriction enzyme became political (see later) but the premise was that the minisatellite had a limited base sequence combination and thus would be resistant to restriction digestion by a high frequency (4-base recognition) enzyme if an appropriate cleavage recognition sequence was chosen. In such a manner the number of repeats of the minisatellite sequence would principally determine the length of the DNA fragment generated as there would only be a relatively smaller extent of flanking sequence. The ideal analytical system for DNA fingerprint minisatellite fragments would have been an electrophoretic system that could resolve to within ~one repeat length (e.g. ~10bp) in the 3–20kb range. As such a system does not exist the size separations required a 0.7% agarose gel system, but the accuracy of size measurement was limited to within a few percent of molecular weight. Thus a failing of the system was that two closely sized fragments, perhaps differing by just one repeat, could not be distinguished. Analytical statistical methods had to be evolved to address this situation.

The DNA fragments were blotted onto a nylon membrane and then probed with one or more multilocus probes. These probes were initially radiolabeled but later became available in chemiluminescent format. The use of multilocus probes at low stringency conditions must be under strictly controlled conditions: a change in concentration of the probe effects the relative intensity of its hybridising to the bound minisatellite sequences, as does a slight change in the hybridisation temperature or the length of hybridisation time, or the extent and nature of the post-hybridisation washes. Full details of a successful procedure can be found in a report by Smith et al. (1990).

The resulting DNA fingerprint pattern is a complex series of bands of differing intensities on an autoradiograph. Despite several systems being developed for image capture and computerised image analysis, inevitably the trained human eye was the mainstay of DNA fingerprint analysis.

4.4 DNA Fingerprinting – The Applications

In fairly quick succession papers mapped out the applications of DNA fingerprinting. The applications ranged from the obvious application of forensic casework (Gill et al. 1985), and paternity and immigration family relationship testing (Jeffreys et al. 1985c, 1986), to its use in animal studies, e.g. for dogs (Morton et al. 1987), birds (Burke and Bruford 1987) and fish (Taggart and Ferguson 1990). Intriguingly, the human sequence probes 33.6 and 33.15 detected fingerprint patterns in all vertebrates tested. Variant multilocus probes based on other sequences allowed applications to extend to plants (Nybom and Schaal 1990).

The discovery of DNA fingerprinting was hailed as a revolutionary technique in identification and was immediately sought after by the police to provide certain identification in casework. The first court case in which DNA fingerprinting evidence was central to a prosecution case was in Bristol, UK, in November 1987. In this case, DNA fingerprints provided the link between a burglary and a rape. The analysis was simple in that DNA obtained from the sperm swabbed from the rape victim had the identical DNA fingerprint pattern to that obtained from a sample of the accused burglar. The samples were run side-by-side in a gel analysis. The observed match clearly needed a statistical weight. Jeffreys had established that within the analysed size range ($\sim$ 4–20 kb) the investigation of random pairing of unrelated DNA fingerprint patterns showed that a band of a certain size range had a small chance of being matched by a band in another person (Jeffreys et al. 1985b). In subsequent studies, this statistic was seen to be conservative (Jeffreys et al. 1991a). Thus a 10-band DNA fingerprinting pattern would only be expected to be matched by one in 4^{10} ($\sim$1 in a million) people. In this first case, in the absence of DNA evidence, the link between the burglary and the rape would not have been established, as other evidence was circumstantial. Since that time hundreds of cases around the world have used DNA evidence, but only a relatively small number have used multilocus probes as the technology was relatively refractive to the normal nature of forensic samples (see below).

The application of DNA fingerprinting to the analysis of relationships was equally well received. In 1988, DNA fingerprinting results were fully accepted by the Home Office as proof of relationships in immigration appeal cases (Home Office London 1988). In 1989 the Home Office gave official recognition to the role of DNA fingerprinting in paternity casework by amending the blood test regulations to allow the granting of official registered tester status to private DNA testing laboratories. In relationship analyses the patterns of mother, child and alleged father need to be analysed side-by-side on the same gel. In this manner bands in the child's DNA fingerprint that have an equivalent with bands in the mother's pattern can be identified. In paternity casework it is then assumed that these bands could have been inherited by the child from the mother as maternity is not in dispute. The remaining unmatched bands in the child's pattern must have been inherited from the true biological father of the child. A visual inspection of the alleged father's DNA fingerprint would rapidly identify whether he has these "paternal" bands in his profile. In this casework it was found that approximately seven or eight "paternal" bands would be found in each of the two multilocus probe results routinely used (Jeffreys et al. 1991a). If these were not matched between man and child, then paternity was excluded. If the bands were matched by bands in the alleged father then this was very strong evidence for parentage. An unrelated man would be expected to have only a quarter chance of matching each band. With 15 paternal bands identified from two multilocus probe tests it can be seen from the statistics mentioned above that it is most unlikely that the man's DNA fingerprint matches by chance.

Immigration casework is far more complicated as the analyst cannot assume maternity, or there may only be one parent available. In such a case the analysis of relationships depends on the level of band sharing between claimed parents and children and not on specific band inheritance. As the analyst is only sampling part of the total DNA fingerprint, the band share between true parents and children will fall into a distribution around 62.5% band share. This curious figure relates to the fact that 50% of the DNA fingerprint bands will be inherited, and 25% of the remaining 50% will match by chance; hence the 62.5% band share. Unfortunately, any such distribution will overlap to some extent with the distribution of band share expected for second-degree relationships (such as uncle–nephew) which are distributed around 44% band share. Thus the relative likelihood of relationships needed to be calculated given an observed band share. This complexity was clarified when necessary by the use of additional DNA profiling tests (see below).

Despite the complexity, the ability to assess the strength of relationships opened a whole new vista on the analysis of animal behaviour and population structures. Subjects such as parentage issues in dogs, infidelity in house sparrows, and the genetic diversity of wales became suitable for molecular analysis for the first time. Conferences were organised to share the breadth of applications of DNA fingerprinting.

In the laboratory the applications ranged from the monitoring of bone marrow transplants (Min et al. 1988), the analysis of genetic similarity in inbred mouse strains (Jeffreys et al. 1987), to the authentication of tissue culture cell lines (Thacker et al. 1988).

4.5 An Explosion of DNA Fingerprinting Sequences

The utility of the Jeffreys' minisatellite markers prompted scientists the world over to hunt for their own minisatellites. This search was prompted by the observation that Jeffreys' probes worked to varying degrees in different animal species: not unsurprising, a "human" origin minisatellite sequence would find fewer homologous sequences as the evolutionary distance of the animal under investigation increased. Hence the probes 33.6 and 33.15 only showed a small number of DNA fingerprint bands in DNA of the duckbill platypus (pers. observ.).

The search for alternative minisatellite sequences was also prompted by the patenting of the Jeffreys' sequences such that licenses and royalties would be required for any commercial exploitation.

Two principal multilocus systems were developed which were widely applied. Vassart identified a segment of sequence in the bacteriophage M13 that could generate fingerprints on human, animal and plant DNA (Vassart et al. 1987). Epplen and co-workers hit upon the use of simple-triplet repeat hybridisation probes to generate equally useful multilocus probes (Peters et al. 1991).

The tremendous interest in minisatellite sequences for DNA fingerprinting was overtaken in the longer term by the sweep of polymerase chain reaction methodologies which were not readily applicable to the large DNA fragments studied in DNA fingerprinting. However, within a few years of the discovery, it was evident that DNA fingerprinting had shortcomings for the pre-eminent applications of forensics and paternity testing that would limit the longevity of its importance.

4.6 Shortcomings of Multilocus Probes

Multilocus probe (MLP) protocols required a number of non-robust laboratory procedures. Of most practical importance were those of probe labelling, hybridisation and autoradiography, as these directly affected the presence or absence of many DNA fingerprinting bands. The presence of any one particular band could affect results that as evidence might lead to criminal convictions, or prove paternity. MLPs were used under low stringency hybridisation conditions whereby the probe would bind to a range of minisatellite sequences in the sample DNA which had varying degrees of homology with the probe sequence. This stringency was extremely difficult to make exactly reproducible. A slight shift in hybridisation conditions such as temperature, buffer salts, or incubation time could affect probe binding to the continuum of minisatellite homologies: this was also possible for changes in post-hybridisation washes. Further, the MLPs were originally radiolabelled by nick translation which potentially created a variable spectrum of probe fragments which would have differing affinities in low stringency conditions. The final exposure of the ^{32}P-probed DNA to X-ray film often involved holding the film at $-80\,°C$ to enhance and speed up the autoradiographic process. This again had to be as reproducible as possible to ensure comparability of results.

Commercially available MLPs significantly improved these problems as they were manufactured as chemiluminescently labelled oligonucleotides. These probes had simpler and shorter hybridisation protocols and were themselves purified probe entities. The use of known control samples with defined DNA fingerprint patterns also enabled greater assurance of reproducibility.

The greatest certainty for analysis was achieved by analysing all the samples to be compared in the same gel analysis when possible, such that any subsequent vagaries in probing conditions equally affected all samples in the case. However, this was not often practicable in cases were a suspect may not have been found, or where previously analysed samples needed to have their results compared with new crimes.

For forensic casework one of the biggest problems for the general utility of DNA fingerprinting was the requirement for microgram quantities of high molecular weight DNA to be obtained from the crime samples. Both the quantity and quality was often not available. Further unequal quantities or qualities of DNA, say from a comparison of a suspect's blood sample with a trace of sperm from a vaginal swab, could also produce different numbers of bands from the same individual.

Nonetheless, DNA fingerprinting evidence was submitted in many cases in the late 1980s. It was initially received with little challenge, there being limited DNA expertise in the ranks of the defence counsels. However, with time, the evidence became subject to critical challenge, not the least being against the "astronomical" certainties placed against what, in effect, was a simple pattern match. Even in poorly visible patterns, resulting from limiting quantities of DNA, a few matching bands gave strong numerical weight to a claimed linkage between samples. Of particular concern was the fact that the DNA bands in a DNA fingerprint that were claimed to match could not be readily assigned to a particular part of the human genome. Thus DNA bands said to match could not actually be proved to derive from the exact same minisatellite.

Given all these shortcomings it was not surprising to find that the directly related "DNA Profiling" technique, which was developing in parallel to the MLP methodology, rapidly came to the fore in forensic and paternity casework.

4.7 Single Locus Probes – DNA Profiling

In 1986 and 1987, in parallel with the developments in DNA fingerprinting, Jeffreys and others were isolating and studying individual minisatellite regions in greater detail (Nakamura et al. 1987; Wong et al. 1987). Several such loci came to pre-eminence as the minisatellites involved were found to be highly variable in their repeat number variation and thus highly informative loci for identification tests in their own right. However, any one such single locus did not contain sufficient discrimination to uniquely identify an individual, but was capable perhaps of providing a characterisation shared by about only one in fifty people. Probes developed for these loci were termed single locus probes (SLPs or VNTR probes) to distinguish them from the MLPs. The resulting two-band pattern they produced from DNA samples was termed a DNA profile, using the word profile to mean an outline, not the full portrait.

Single locus probes had an immediate set of properties that advocated their use in forensic and paternity casework. They detected a unique DNA minisatellite species under robust highly stringent hybridisation conditions. Thus the resulting DNA profile directly related to a defined human DNA region and did so in all samples examined. Thus comparable sized DNA fragments in different samples equated to identical DNA.

The DNA profiles obtained were simple two-band patterns, each band representing one of the two copies of DNA present in an individual. This was far simpler for comprehension by lay juries and for those involved in paternity applications. In this latter case one of the two bands present in a child's DNA profile represents the DNA inherited from the mother, such that the remaining band must match with one of the two in the alleged father if he is the true biological father of the child.

Importantly, the use of stringent hybridisation conditions, such that the labelled probe now located to a single DNA location, greatly enhanced the detection sensitivity of the technology. DNA profile tests only required approximately a twentieth of the DNA used for DNA fingerprints, which was of great importance to forensic casework. Using less sample DNA also greatly diminished the background of nonspecific hybridisation that is inherent in probing work, thus providing for clear two-band DNA tracks.

The simple clear profile bands allowed for image capture and computerised assessment of band positions. This technology was of great importance as it eliminated any consideration of subjective analyses and produced numerical sizing of DNA bands for comparison between samples. This was also extremely important as it allowed DNA profiles to be compared directly from samples run on different gels on different days, even, in principle, different laboratories. The correct use of control samples and molecular weight standards was essential for such comparisons to be valid and became the focus of much courtroom debate (see lessons from the Castro case).

The statistical assessment of the significance of a DNA match also became easier, at least until court lawyers started to get involved. In principle it was now possible to screen a representative number of people from a distinct population group and record the allelic variation observed. From this the investigator could determine an expected frequency for a particular allele in the population. Thus the statistics associated with the DNA evidence became directly related to the observed DNA type present and was not derived from a formula that treated all DNA bands equally (as used for DNA fingerprinting).

The laboratory methodology employed for DNA profile analysis was akin to that used for DNA fingerprints. High molecular weight DNA still needed to be isolated and restricted. The DNA fragments were still sorted by the same gel systems used and the DNA patterns were blotted onto nylon membranes as before. Importantly now though, as one SLP test was insufficient for identification purposes, the nylon-bound DNA was subjected to sequential probing with different SLPs. Each SLP analysed an independent minisatellite in human DNA and thus provided a DNA typing result that could be combined with that of other SLPs so as to accumulate sufficient evidence to effectively individualise the DNA of the sample involved. Each probing inevitably reduced the utility of the nylon-bound DNA for the next probe, such that, for samples with limited DNA to start with, only a few probe results could be obtained.

The computerisation and hence intercomparability of DNA profile results, coupled with the ability to define the statistical weight of any DNA profile band, promised a hey-day in international forensic collaboration and exchange of DNA methods, data, probes, databases, etc. Sadly, however, Europe and the USA decided to opt for different standards in one aspect of the methodology that greatly reduced the global potential of this technology.

4.8 Europe and the USA Disagree

Jeffreys had chosen Hinf I as the restriction enzyme to digest genomic DNA for his MLP studies and this enzyme was also found to be suitable for the SLP probes his laboratory subsequently derived. In the UK therefore the combination of probes and restriction enzyme developed by Jeffreys became the profiling combination used by the Home Office forensic scientists and Cellmark Diagnostics, the ICI company set up to offer DNA fingerprinting applications. This combination was taken up in turn by the majority of European forensic DNA laboratories. In the USA, two companies taking a lead in this field were Lifecodes and Collaborative Research and they both used the Pst I enzyme with their probes which they recommended. In 1988 the FBI had to decide which system to adopt, and rather than showing any preference to either side decided upon a different enzyme called Hae III. The consequent differences in choice of restriction enzyme led to a level of incompatibility of SLP data between the USA and Europe as the two enzymes tended to cut within the repeating units of each other's preferred probes. As the technology was adopted around the world the differing forensic allegiances led to a patchwork of methodologies being used.

Over the next few years two probes were commonly used by the American and European systems, MS 1 (D1S7) and YNH24 (D2S44), as both were suitable for use with either Hae III or Hinf I. However, the majority of the other probes used in the two systems were different. In a later paper Jeffreys (Wong et al. 1987) changed to using the enzyme Alu I (cutting AGCT) rather than either Hinf I or Hae III, as their recognition cut site sequences (GANTC and GGCC) were both commonly found in the G-rich minisatellites and thus potentially limited the applicability of new highly informative minisatellites that were being discovered. In the UK the use of Alu I was taken up by a new DNA-profiling company called University Diagnostics, which could then use both European and American probe systems. However, this approach was not really adopted elsewhere. In practice the choice of probes and restriction enzymes was just one of many challenges facing the implementation of DNA profiling as the results were now being entered into court as evidence. Every facet of the DNA methodology was now facing the full scrutiny of the judicial system, especially in the USA.

4.9 Lessons from the Castro Case

The DNA evidence for the Castro case arguably changed the face of DNA evidence analysis and its court presentation. In what otherwise would not have been a notably significant double murder in New York, this case gained notoriety as six different areas of the DNA evidence were rigorously challenged and found to be wanting.

In February 1987 Vilma Ponce and her daughter were stabbed to death in their Bronx apartment. A tip-off led the police to arrest a neighbourhood handyman called Jose Castro. The key piece of forensic evidence for the prosecution case was the finding of a small bloodstain on the watch of Castro that was sent for DNA typing. The DNA studies were performed by Lifecodes, a private American genetic testing company. DNA (0.5 µg) was extracted from the bloodstain and was compared with that of the victims. Three SLP probes were used (D2S44, D17S79, DXYS14) along with a Y chromosome probe (DYZ1). Lifecodes reported that the pattern of DNA bands found on the watch matched that of the mother: three bands matched for DXYS14, one for D2S44 and two for D17S79. A frequency of 1 in 100,000,000 was reported for such a match. The company reported no difficulties or ambiguities in their analysis.

A defence analysis of the DXYS14 result, however, questioned the findings reported. Five bands were noted in the sample from the watch, whilst only three were reported matching the three in the mother. In testimony, the lead scientist agreed that there were extra non-matching bands, but asserted they could be discounted as being non-human contamination. The scientist argued that the two extra bands had a different intensity to the reported three and this was unexpected, as the intensity of SLP bands decreases consistently with reduced fragment size. However, an examination of laboratory data indicated that an ironclad intensity profile could not be established.

The company sized the DNA fragments in the case by an objective digitising system that they claimed indicated that the laboratory had an operational 0.6% deviation on a given band size. The analysis rule for the company indicated that two bands that fell within three standard deviations of each other are considered indistinguishable and are reported only as the same average size. The actual band size comparison (watch band vs. mother band) for D2S44 was a range of 3.06 SDs and for one of the D17S79 bands was 3.66 SDs. Under cross-examination the lead scientist said these results did not adhere to the objective threshold, the decision being based on a subjective visual comparison.

The company reported that the Y probe result indicated that the blood on the watch was female as no hybridisation band was observed. However, the defence examination revealed that the control sample in the analysis also gave no hybridisation, indicating that perhaps the hybridisation step had not worked at all. The source of the control sample was thus key to the interpretation of this result. However, the company changed its tune through a series of responses; firstly, the control was indicated to be from the female cell culture line HeLa, but this was contradicted by a technician who said it was a male technician's DNA, the lack of hybridisation was then explained by the technician having a rare short Y chromosome. However, this was then shown to be incorrect and the sample actually came from a female technician traced not by notes but by comparing DNA results from other probes on the control with records of the laboratory staff. The outcome of all this was that, with no positive control, the Y probe result could not be validated.

With the D2S44 probe only one band was observed at a size of 10.25 kb, yet 90% of the alleles with this probe were actually larger in size. To rule out DNA degradation within the watch sample, which could effectively remove a larger fragment that might exclude a match, the company reprobed the DNA with an Alu I probe that showed bands up to 23 kb. However, the defence showed that a technician had misread the size marker ladder such that the Alu I ladder only measured up to 9.8 kb. Thus DNA degradation could not be ruled out.

To explain their interpretation of the SLP results, the prosecution scientists invoked four separate instances of contamination in their probe stocks. The company nonetheless continued to use the probes after the contamination was noted, although no precise records were kept of the use of the probes.

Finally, the company did not use the same matching rule for declaring a match and valuing it. The company used ±3 SDs to declare a match, but only applied a window of ±2/3 SDs to calculate the value from the reference database. Such a method would approximately over-value the significance of each band match by a factor 4.5-fold.

The Castro case became internationally renown when an article by Lander, who had played a pivotal role for the defence in challenging the evidence, was published in *Nature* (Lander 1989). Lander recounted the blow-by-blow account of the case and its presentation in court and in doing so provided a template of issues by which a DNA profiling case should be considered.

Lander highlighted the need for standards both in the laboratory methodology, the analysis and the reference databases used. The setting and monitoring of these standards became the focus of several international committees, but, importantly, most working laboratories had the majority of the requirements already in place. Thus the Castro case served to reinforce attention to correct detail and brought DNA evidence off its pedestal. In the following years the fine detail of DNA casework in the areas highlighted by Castro were constantly challenged, but it was actually only the rare case in which the DNA evidence was found inappropriate and unacceptable. In 1994 Lander co-authored a paper with Budowle, Head of DNA at the FBI, to again address the issue of DNA evidence and review in particular aspects of population statistics. This time, however, it was acknowledged that the disputed issues were essentially properly addressed by current procedures (Lander and Budowle 1994).

4.10 Impact of PCR

SLP profiling held pre-eminence as the method of DNA typing from 1989 to 1994. However, at a time when the technology was well bedded in forensic casework and might have been expected to hold sway for years to come, a new generation of tests took over completely. The new generation of DNA profile tests came about through the invention of the polymerase chain reaction (PCR; Mullis 1990).

PCR has revolutionised DNA investigative science in all fields of molecular biology by enabling any DNA sequence to be analysed by the simple expedient of designing primers that flank the sequence of interest (see Chap 5). For forensic science it enabled scientists to develop tests using PCR that addressed three major shortcomings of SLP analysis; the quantity of DNA required to be analysed, the time taken to analyse the DNA and the accuracy of determining exactly what DNA type(s) are present in a sample. The latter capability has in turn brought about the confidence to create major databases of DNA identities (see Sect 4.13).

Forensic casework rarely handles "ideal" samples from the scene of a crime as time and the environment work to degrade the quality and quantity of the DNA present. SLP alleles range in size from approximately 500 bp up to something in the order of 20,000 bp. Any environmental agent leading to the fracture or degradation of DNA, such as microbial action or UV, inevitably will reduce the size of the DNA fragments in the sample. This, in turn, leads to decreasing amounts of full-length alleles of each SLP loci for analysis. PCR allows one to investigate small lengths of DNA, say from 50 bp upward, because it can manufacture sufficient copies of such sequences so that they can be detected. This is a difficult challenge for a probe-based analysis where the detected signal reflects the size and quantity of the probe-binding sequence present to start with. Thus, by choosing an appropriate DNA identity test system able to provide distinguishing DNA alleles within small fragments, a PCR-based system can be informative from minute quantities even of degraded DNA, potentially from just a single cell (Jeffreys et al. 1988). This has opened up casework that could not be analysed by SLP methodology.

The SLP probing system took 4 to 6 weeks to process a sample through a number of probes sufficient to give a unique identification. Any failure in any of the probes could delay the process further, as could waiting for longer exposures of small quantities of DNA. In contrast, the PCR method can generate millions of copies of the allele to be analysed in a couple of hours from samples from which DNA has been rapidly extracted. Rapid extraction is possible because the DNA no longer has to be isolated gently to obtain maximum lengths of DNA, and, equally, the DNA does not need to be of great purity as PCR just needs a subset of DNA molecules suitable to be copied (Walsh et al. 1992). Thus, in theory, results from samples can be generated within the day. In any police investigation time is of essence and PCR offered a radical improvement that could not be matched by probing methodologies.

As described above minisatellite alleles encompassed such a wide range of possible DNA sizes that the agarose gel systems employed to cope with such possibilities suffered from a consequent lack of size resolution of the DNA fragments involved. Most SLP allele sizes had a resolution of the order of ±2.8% of determined size. Thus an SLP allele of 10,000 bp could encompass a range of 500 bp, which potentially could include 20 or so totally different DNA alleles (assuming a repeat length of 25 bp) which would be defined as being indistinguishable. Whilst arguably this inaccuracy was taken into account in the statistical weight applied to a "match" between fragments, it clearly could miss the observation of an exclusion. PCR, by allowing the analysis of small DNA fragments, has meant that DNA fragments can now be analysed in sequencing gels reading to single base-pair accuracy. This accuracy has radically simplified the debate as to when a match can be called: either the same DNA fragment base size is detected or it is not the same DNA fragment.

PCR was invented in 1987, but the recognition and development of its value to forensic science, as well as the validation of potentially radically different DNA identity tests, took some time to settle down to a universally applied methodology. The presently applied technology is called short tandem repeat (STR) profiling and will be addressed in Section 4.12. STR profiling came into service in 1994 in the UK. In the interim following the invention of PCR a number of alternative DNA identification tests came to the fore briefly. A first port of call was to evaluate the applicability of PCR to the pre-existing SLP allele system. Jeffreys examined PCR amplification of the SLP alleles to achieve results from potentially multiplexed SLP loci, analysed simultaneously (Jeffreys et al. 1988). However, the rapidity of the PCR analysis from potentially smaller samples was countered by the need to keep the number of PCR amplification cycles low to stop mispriming artifacts. Thus, this approach still required the same probing methods and faced the same gel resolution issues and thus the forensic community did not take up this method.

However, the study of smaller minisatellite systems with allelic variants limited to the ~1000-bp range presented potential advantages. Several such systems, known as AMFLPs (amplified fragment length polymorphisms) were developed to benefit from PCR and sharper resolution in acrylamide gel systems. These were used alongside SLP profiling through to the late-1990s (Budowle et al. 1991).

A system which was taken up on an international basis, the DQalpha system, took a radically different approach to identification and moved away from repeat sequence DNA analysis to focus instead on the hypervariable sequence within the major histocompatibility locus (Saiki et al. 1989). Roche has manufactured a family of kits based on multiple points of analysis of common polymorphisms with this DNA region. In the USA, in particular, these systems were used in combination with AMFLPs and SLPS to give additional DNA evidence in a case. These kits did not have the discriminating capability of the minisatellite systems, neither did the emerging STR profiling systems based on microsatellite loci.

Of particular novelty and elegance was another invention by Jeffreys, that of digital DNA typing. This merits special attention as Jeffreys used PCR to look inside the actual mechanism of repeat sequence variation in minisatellites and created a method that could potentially uniquely identify an individual from just one or two minisatellites.

4.11 Digital DNA Typing

The analysis of minisatellite sequence variation for identity analysis had focused on the variations of the DNA fragment length caused by changes in the number of repeats of the minisatellite sequence. Jeffreys utilised PCR to investigate the actual molecular mechanisms underlying the repeat length changes and in doing so uncovered tremendous variation of sequence structure actually within a minisatellite. Jeffreys discovered that the actual repeating sequence was not always identical and that certain variants had become established along the array of repeats. Characterisation of many alleles at the MS32 locus revealed a polar mutation/recombination mechanism underlying the variation in minisatellite repeats. This mechanism was seen to generate a random evolution of the order of the minisatellite repeat variants. The variation in the order of repeat types within an individual's two alleles could be easily analysed on a sequencing gel and converted to a simple numerical code. Jeffreys coined the term digital DNA typing for this new identification technology (Jeffreys et al. 1991b). The coding of an individual's repeats at just one locus like MS32 provided near unique identification.

One difficulty for the use of digital DNA typing, also known as minisatellite variant repeat PCR (MVR-PCR), was the complicated analysis that would result from samples containing the DNA of more than one individual. In this case the analysis would super-impose the two individuals' repeat variants to give a hybrid pattern. To overcome this problem MVR-PCR was complemented by the discovery and use of polymorphisms in the DNA sequences flanking the minisatellite so as to create a potential series of haplotype codes for analysis (Monckton et al. 1993). However, this elegant methodology was technically complicated and its interpretation was more difficult than the STR profiling technology that was now taking centre stage. Thus MVR-PCR remains a tool for the investigation of mutational mechanisms in minisatellites, but has had little application in forensic casework.

4.12 Short Tandem Repeat Profiling

Short tandem repeats (STRs) are a simple cousin to the variable number tandem repeats found in minisatellites, but consist of short repeating sequence motifs of only 2–6 bp. STR loci are thus known as microsatellites. It is estimated that the human genome has 500,000 STR loci; however, most are simple dimeric repeats and the repeats of 3 or 4 bp are numbered in the thousands. The attraction of the microsatellite systems for DNA identification analysis is that, in principle, all the logic, history and experience of minisatellite analyses can be transposed directly into a repeat variation that will be contained within a small fragment size that is ideally suited for PCR methodology.

The simple and abundant dimeric repeat sequences were initially analysed for their suitability (Weber and May 1989). Whilst successfully amplified, the 2-bp repeats were found to suffer from some misprocessing in PCR by the Taq polymerase which provides the DNA amplification. As a consequence an allele was observed with a family of sub-alleles which were either one, two or three, etc., repeats shorter or longer that the original (Litt et al. 1993) Thus, given that any locus had two alleles, each with their amplification artifacts that would be super-imposed, definitive analysis suitable for court scrutiny was not readily achievable. Further, the 2-bp repeat loci investigated appeared to have only a limited number of allelic variants present in the population and thus did not provide a substantial differential between an individual and the population.

However, the STRs based on 3- and 4-bp repeat sequences did not appear to suffer from this artifact to any significant extent and, at the same time, had a larger range of allele variants. In the UK, the Forensic Science Service devised a panel of four STR loci which had 4-bp repeats and represented a new testing package encompassing all the advantages of PCR methodology (Kimpton et al. 1994). In this first model STR system the four loci could be analysed simultaneously by the tagging of the primers for each loci with different fluorescent tags. The DNA alleles were identified during electrophoresis by equipment that scanned the migrating DNA with a laser that activated the tags. An additional fluorescent tag was applied to molecular weight standards that were run within every track so as to provide extremely accurate size analysis in combination with 6% denaturing polyacrylamide gels. The STR quadraplex could work efficiently with samples down to ~100 pg of DNA, a thousandth of that required for SLP probing. The four loci alleles were contained within a small size range of DNA fragments (130–240 bp) allowing the quadraplex to be applicable to real forensic casework samples. The complete analysis of these microsatellite systems only took a matter of hours once the DNA had been extracted from the sample.

The quadraplex was implemented into forensic casework in the UK in 1994 and rapidly replaced the SLP methodology that had long held sway. Of particular value to juries, who had long suffered with definitions of kilobases and standard deviations, was the ability to identify specific repeat alleles for each locus. Thus, for instance, the DNA evidence for the alleles of the STR test HUMTHO1 could be simply reported as, say, 6,7 (or other numerical allele designations), such that the jury would be looking for a match against another 6,7 combination. The accuracy and simplicity of the use of STRs for allele designation provided a ready platform for the computer databasing of STR profiles. This in turn laid the groundwork for the UK to introduce the first National DNA Database.

4.13 The UK National DNA Database

In 1995 the UK Government enacted legislation to create the first National DNA Database (NDNADB). This empowered the police to take swabs or hair roots from anyone charged with an indictable crime and for the STR profile to be prepared from the sample and entered into a national computer system. The profile would be stored if the accused was found guilty of the crime and would be searched against the profiles of all unsolved crimes. To minimise inadvertent matching of profiles between unrelated individuals, it was necessary to beef up the discrimination of the STR profiling system from four to six loci (Sparkes et al 1996). This system, known as the second-generation mix (SGM), is claimed to distinguish on average one person amongst 56 million, the approximate population of the UK. The six loci chosen were THO1, VWA, FGA, D8S1179, D18S51 and D21S11. In addition, a sex test, known as amelogenin, is included. In 1999 there were approximately 500,000 samples and 50,000 crime stains typed by their SGM profile on the NDNABD. Presently, 400–700 matches are identified per week.

The growth in the use of the NDNABD has increased concern that adventitious matches may be found with the SGM, as a profile of common allele types could occur as frequently as 1 in 10,000 individuals. The SGM was therefore complemented in 1999 by four additional loci to create a multiplex of 10 loci with a chance match now claimed to be of the order of 1 in a 1000 million. This new multiplex is called SGM Plus and came into use in the autumn of 1999 for both casework and the NDNADB. Existing SGM profiles will remain on the database.

The SGM has been well received on an international basis. A recent Interpol DNA expert group has set a European standard using the SGM with one additional STR locus. These are routinely used in forensic casework throughout Europe and have become core STR loci for European DNA analytical systems. In Europe, the drive from national DNA databases is much more muted, and emphasises the databasing only of convicted felons.

In 1998 the USA introduced an STR-based index comprising 13 STR loci including the 6 SGM loci. This federal database, known as CODIS, is now being used by a number of states and also includes additional testing data from SLPs and AMFLPs.

4.14 STR Profiling Applications

In theory, the spectrum of applications for microsatellites should become similar to that presently served by minisatellites as users take advantage of the cost and speed improvements offered by a PCR-based test.

Paternity testing has predominantly shifted over to STR profiling and has been achieved by replacing 4–6 SLP probe tests with ~12 STR tests (Thomson et al. 1999). More loci have been required because the STR loci have a limited number of allelic variants and are thus, test for test, less informative. This has the added benefit that discrimination between closely related individuals is improved by increasing the number of test systems challenging the claimed relationship, but has a disadvantage in that with more systems there is an increased likelihood that mutation events will be observed. As a consequence, if one is considering a panel of perhaps 16 STR loci, the observation of one or two excluding alleles in a claimed relationship cannot be assumed to exclude the relationship. In such cases one must consider statistically the relative chance of mutations in a true relationship, versus an unrelated pairing matching on all but one or two STR loci tests. These considerations have been central to the development of STR profiling tests for the analysis of complex familial relationships (Thomson et al. 2000).

Veterinary identity and pedigree applications for STRs have already progressed to commercially available STR multiplex kits from Perkin Elmer. There are already such systems available for cattle, horses (see, for instance, Bowling et al. 1997) and dogs. These systems provide easy sampling, such as blood pricks or hair roots, because of the PCR basis, and are relatively inexpensive and can be completed rapidly. STR-based databasing of horses, livestock and dogs for identification and pedigree confirmation is already in use, despite this being a novel development as minisatellites were not readily databased and were too expensive as a technology for this application.

One disadvantage of STRs, however, compared with multilocus probes, is that they tend to be species-specific. Thus the human STRs are uninformative in other mammalian DNA samples. This is particularly relevant for an application like cell line typing, for which most laboratories have stocks of human cultures alongside rodent or monkey lines and for which specific subsets of STRs would have to be isolated and characterised. In the avian world, particularly in pet and exotic bird collections, an identification and parentage test would be welcomed that covered all avian species. To date this remains elusive.

4.15 The Future

The scientific journals are full of the next generation of tests, i.e. chip-based hybridisation array technologies. At present, these systems are being engineered to detect single nucleotide polymorphisms (SNPs). The chips are designed to hold arrays of short lengths of possible nucleotide sequence permutations to match gene sequences of interest. The presence of SNPs will be detected by both the presence or absence of hybridisation to the discriminating sequences. Minisatellites and even microsatellites may be very difficult to analyse in such systems, as the DNA is characterised not by the presence or absence of a base pair, but by the presence of different numbers of copies of the same sequence. As this field is driven by pharmaceutical research, and microsatellites are a distraction rather than a focus for that industry, it is probable that these systems will not be rapidly adapted to accommodate microsatellites analyses. Rather it is probable that an identification system based on SNPs will come to the fore.

Because of the incompatibility of STR analyses and gene chip systems, as well as the lack of overlap between SNP and STR results, it is not clear whether the transformation to gene chip technology will be an attractive proposition for sometime. The transition will require a conversion, or new analysis of all historic samples for big initiatives such as national databases, and the cost of these changes must be borne by governments or institutions. Conversely, applications of minisatellites and microsatellites that do not require reference databases could readily utilise SNP-based systems, but these applications may be of insufficient market value to attract chip application companies.

Therefore, for the next 5 years or so, it would appear that microsatellite technology will have a place in identification and relationship applications. However, the era of multilocus and minsatellite probing applications has now come to an end, just 14 hectic years after Jeffreys' invention surprised the world.

References

Bowling AT, Eggleston-Stott ML, Byrns G, Clark RS, Dileanis S, Wictum E (1997) Validation of microsatellite markers for routine horse parentage testing. Anim Genet 28:247–252

Budowle B, Charkraborty R, Guisti AW, Eisenberg AJ, Allen RC (1991) Analysis of the VNTR locus D1S80 by PCR followed by high resolution PAGE. Am J Hum Genet 48:137–144

Burke T, Bruford MW (1987) DNA fingerprinting in birds. Nature 327:149–152

Gill P, Jeffreys AJ, Werrett DJ (1985) Forensic applications of DNA fingerprints. Nature 318: 577–579

Home Office (1988) DNA profiling in immigration casework: report of a pilot trial by the Home Office and Foreign and Commonwealth Office. Home Office, London

Jeffreys AJ, Wilson V, Thein SL (1985a) Hypervariable "minisatellite" regions in human DNA. Nature 314:67–73

Jeffreys AJ, Wilson V, Thein SL (1985b) Individual-specific fingerprints of human DNA. Nature 316:76–79

Jeffreys AJ, Brookfield JFY, Semeonoff RS (1985c) Positive identification of an immigration testcase using human DNA fingerprints. Nature 317:818–819

Jeffreys AJ, Wilson V, Thein SL, Weatherall DJ, Ponder BAJ (1986) DNA fingerprints and segregation analysis of multiple markers in human pedigrees. Am J Hum Genet 39:11–24

Jeffreys AJ, Wilson V, Kelly R, Taylor BA, Bulfield G (1987) Mouse DNA fingerprints: analysis of chromosome localization and germ-line stability of hypervariable loci in recombinant inbred strains. Nucleic Acids Res 15:2823–2836

Jeffreys AJ, Wilson V, Neumann R, Keyte J (1988) Amplification of human minisatellites by the polymerase chain reaction: towards DNA fingerprinting of single cells. Nucleic Acids Res 16:10953–10971

Jeffreys AJ, Turner M, Debenham PG (1991a) The efficiency of multilocus DNA fingerprint probes for individualization and establishment of family relationships, determined from extensive casework. Am J Hum Genet 48:824–840

Jeffreys AJ, MacLeod A, Tamaki K, Neil DL, Monckton DG (1991b) Minisatellite repeat coding as a digital DNA approach to DNA typing. Nature 354:204–209

Kimpton C, Fisher D, Watson S, Adams M, Urquhart A, Lygo J, Gill P (1994) Evaluation of an automated DNA profiling system employing multiplex amplification of four tetrameric STR loci. Int J Legal Med 106:302–311

Lander E (1989) DNA fingerprinting on trial. Nature 339:501–505

Lander E, Budowle B (1994) DNA fingerprinting dispute laid to rest. Nature 371:735–738

Litt M, Hauge X, Sharma V (1993) Shadow bands seen when typing polymorphic dinucleotide repeats: some causes and cures. Biotechniques 15:280–284

Min GL, Hibbin J, Arthur C, Apperley J, Jeffreys A, Goldman J (1988) Use of minisatellite DNA probes for recognition and characterization of relapse after allogenic bone marrow transplantation. Br J Haematol 68:195–201

Monckton DG, Tamaki K, MacLeod A, Neil DL, Jeffreys AJ (1993) Allele-specific MVR-PCR analysis at minisatellite D1S8. Hum Mol Genet 2:513–519

Morton DB, Yaxley RE, Patel I, Jeffreys AJ, Howes SJ, Debenham PG (1987) Use of DNA fingerprinting analysis in identification of this sire. Vet Rec 121:592–593

Mullis KB (1990) The unusual origin of the polymerase chain reaction. Sci Am 262:36–43

Nakamura Y, Leppert M, O'Connell P, Wolff R, Holm T, Culver M, Martin C (1987) Variable number of tandem repeat (VNTR) markers for human gene mapping. Science 235:1616–1622

Nyborn H, Schaal BA (1990) DNA "fingerprints" applied to paternity analysis in apples. Theor Appl Genet 79:763–768

Peters C, Schneider V, Epplen JT, Poche H (1991) Individual-specific DNA fingerprints in man using the oligonucleotide probe $(GTG)_5/(CAC)_5$. Eur J Clin Chem Clin Biochem 29:321–325

Saiki RK, Walsh PS, Levenson CH, Erlich HA (1989) Genetic analysis of amplified DNA with immobilized sequence-specific oligonucleotide probes. Proc Natl Acad Sci USA 86:6230–6234

Smith JC, Newton CR, Alves A, Anwar R, Jenner D, Markham AF (1990) Highly polymorphic minisatellite DNA probes. Further evaluation for individual identification and paternity testing. J Forensic Sci Soc 30:3–18

Sparkes R, Kimpton C, Watson S, Oldroyd N, Clayton T, Barnett L, Arnold J, Thompson C, Hale R, Chapman J, Urquhart A, Gill P (1996) The validation of a 7-locus multiplex STR test for use in forensic casework (1). Int J Legal Med 109:186–194

Stedman R (1983) Bloodgroup frequencies of immigrant and indigenous populations from south east England. J Forensic Sci Soc 25:95–134

Taggart JB, Ferguson A (1990) Minisatellite DNA fingerprints of salmon fishes. Anim Genet 21:377–389

Thacker J, Webb MBF, Debenham PG (1988) Fingerprinting cell lines: use of human hypervariable DNA probes to characterize mammalian cell cultures. Somat Cell Mol Genet 14:519–525

Thomson JA, Pilotti V, Stevens P, Ayres KL, Debenham PG (1999) Validation of short tandem repeat analysis for the investigation of cases of disputed paternity. Forensic Sci Int 100:1–16

Thomson JA, Ayres KL, Pilotti V, Barrett MN, Walker JIH, Debenham PG (2000) Validation of multiplex STR systems for the investigation of parentage and other familial relationships. In: Sensabaugh GF, Lincoln PJ, Olaisen B (eds) Progress in forensic genetics. Elsevier Science, Amsterdam, pp 374–376

Vassart G, Georges M, Monsieur R, Brocas H, Lequarre AS, Christophe D (1987) A sequence in M13 phage detects hypervariable minisatellites cloned from human DNA. Science 235:683–684

Walsh PS, Metzger DA, Higuchi R (1992) Chelex 100 as a medium for simple extraction of DNA for PCR-based typing from forensic material. Biotechniques 10:506–513

Weber JL, May PE (1989) Abundant class of human DNA polymorphisms which can be typed using the polymerase chain reaction. Am J Hum Genet 44:388–396

Wong Z, Wilson V, Patel I, Povey S, Jeffreys AJ (1987) Characterization of a panel of highly variable minisatellites cloned from human DNA. Ann Hum Genet 51:269–288

Wyman A, White R (1980) A highly polymorphic locus in human DNA. Proc Natl Acad Sci USA 77:6754–6758

5 Multiplex Polymerase Chain Reaction and Immobilized Probes: Application to Cardiovascular Disease

SUZANNE CHENG

5.1 Introduction

Cardiovascular disease (CVD) is one of many common diseases for which there appear to be multiple genetic and environmental risk factors that interact with one another during the disease process. Among some families, causative mutations in certain genes have been identified, such as for familial hypertrophic cardiomyopathy, long-QT syndrome, and Marfan syndrome (reviewed by Maron et al. 1998); however, the association of these same genes with common forms of CVD is not fully understood. Extensive epidemiological studies have identified risk factors for disease in the general population, including age and gender, high serum cholesterol levels and hypertension, cigarette smoking, and physical inactivity (reviewed by Pasternak et al. 1996). Familial clustering (ten Kate et al. 1982) and twin studies (Marenberg et al. 1994) indicate that family history is an independent risk factor, particularly for individuals before the age of 55, and that this is not entirely explained by familial aggregation of hyperlipidemia, diabetes mellitus, obesity, and hypertension. Many premature CVD cases remain unexplained (Hoeg 1997), and unknown risk factors may be primarily genetic. These genetic risk factors may include the cumulative effects of common allelic variants; although individual variants may contribute only a modest increase in risk, combinations of particular alleles may confer significantly greater risk for disease. Consequently, efforts to understand CVD in the general population involve the identification of common multilocus genotypes that confer high risk and mechanisms by which lifestyle factors can modulate the expression of a genetic predisposition.

Advances in genome scanning techniques have led to the identification of an increasing number of genes, including many genes implicated in pathways contributing to the development of CVD. These efforts are complemented by association studies to compare affected and unaffected individuals, with the expectation that genetic risk factors will be enriched in frequency among those with disease. Repetitive sequence markers, or microsatellites, associated with each candidate gene may be used for such studies (see Chap. 3). An alternative approach is to utilize single nucleotide polymorphisms (SNPs, reviewed by

Principles and Practice
Molecular Genetic Epidemiology – A Laboratory Perspective
Ian N.M. Day (Ed.)
© Springer-Verlag Berlin Heidelberg 2002

Schafer and Hawkings 1998) that may serve simply as genetic markers, or may in fact be of functional relevance. The many biological pathways leading to CVD implicate hundreds of genes, and understanding genetic predisposition is likely to require studies of several hundred to thousands of individuals with carefully defined clinical symptoms or phenotypes. Each individual must be genotyped for each candidate gene to evaluate that gene's association with disease, in the context of the complex etiology of disease. Assays allowing multiple genetic markers to be queried simultaneously offer significant savings in time, costs, and genetic materials.

The CVD "multiplex" approach described here builds upon two techniques that have greatly facilitated the analysis of genetic variation, the polymerase chain reaction and sequence-specific oligonucleotide probes. Co-amplification of multiple targets and subsequent detection of multiple alleles simultaneously can be adapted to the study of a few genes (tens of markers), as described here, or for thousands of genes, as with a DNA chip (Chap. 4). This approach can be applied to groups of single-gene disorders, to single-gene disorders with many contributing mutations (e.g., cystic fibrosis), as well as to complex diseases with many contributing genes (e.g., CVD). Such assays may also serve as clinical tools, supplementing current diagnostic measurements for complex, common diseases such as cardiovascular disease.

5.2 Multiplex Polymerase Chain Reaction (PCR) Methodology

By amplifying genomic regions of interest, PCR (Mullis and Faloona 1987; Saiki et al. 1988) methods dramatically reduce the amount of genomic material needed for the analysis of genetic variation. A further advance has been to pool multiple PCR primer pairs such that a single reaction enriches for multiple regions of the genome. One of the earliest applications of this approach was to screen six regions of the 2-Mb dystrophin gene; differential mobilities under agarose gel electrophoresis revealed key deletions diagnostic for Duchenne muscular dystrophy (Chamberlain et al. 1988). The success of a multiplex PCR system depends on maximizing the specificity and efficiency of each amplification, for reliable and comparable yields of all products.

5.2.1 Primer Design

As for single-target PCR, primer optimization for a multiplex system is ultimately empirical, but there are several guidelines to follow in selecting primer candidates. Strategies used for single-target PCR primer design must be supplemented to minimize interactions among primers that are likely to occur when multiple primer pairs are pooled. Each primer pair should be tested alone before being combined with other primers, and relative product yields can often be estimated by gel analysis. Further minor adjustments to primer concentrations may be necessary in conjunction with probe optimization.

5.2.1.1 Product Size

To maximize amplification efficiencies, product lengths should be less than 1 kilobase pair (kb), or less than 500 base pairs (bp) if possible. Larger regions of interest should be divided into multiple amplicons. To minimize preferential amplification of shorter fragments at the expense of longer fragments (Walsh et al. 1992), the product size range should be as narrow as possible, for example, at most a 300-bp difference between the shortest and longest PCR products. The practical range will depend on the specific assay, and, if insertions or duplications necessitate the use of larger amplicons, then a somewhat wider product size range may be necessary. Since gel analysis can be used to assess relative yields from multiplex reactions of up to 15 or 20 products, fragment sizes can be chosen to permit gel resolution. As the number of targets increases, however, gel resolution becomes difficult, and optimal amplifications are more likely with narrower product ranges; for these larger multiplexes, probes can be used to confirm target amplification once a primer pair has been incorporated into the primer pool. Occasionally, a PCR product strand will form stable secondary structures that prevent hybridization to an immobilized probe; if a longer probe is unable to compete successfully, detection of this strand may require use of primers that define a shorter amplicon.

5.2.1.2 Primer Length

PCR primers are generally between 18 and 28 nucleotides in length, depending on their base composition. As a group, multiplex PCR primers should have similar melting temperatures; one algorithm for calculating the melting temperature of a primer is described by Rychlik et al. (1990). Higher melting temperatures allow use of higher annealing temperatures during PCR, which generally confer a greater specificity of amplification. However, the practical melting temperature range for any primer pool will depend on the base compositions of the fragments to be amplified.

5.2.1.3 Primer Sequence

One precautionary step critical for multiplex PCR systems is to avoid possible 3'-terminal complementarity that could initiate low molecular weight, target-independent amplifications typically termed "primer-dimer" formation (Li et al. 1990). During PCR setup at room temperature, transient annealing between primers is possible and, in the presence of polymerase activity, sufficient strand synthesis may occur to generate very short products. These short products will act as template molecules during subsequent PCR, and can be so effective in competing for reaction components that amplification of desired targets is significantly reduced. One relatively simple solution is to select primers that share a common, nonpalindromic 3'-terminal sequence. A 40-primer multiplex system for the cystic fibrosis transmembrane conductance regulator gene was developed using common 3'-AA ends; single nucleotides can also be effective at controlling primer-dimer formation (Zangenberg et al. 1999). In the CVD multiplex example described below, allele-specific primers were designed for the codon 112 cys/arg polymorphism of the apoE gene, and these primers consequently established the criteria by which all remaining 3'-ends were selected. As more targets are added to a multiplex, it may not always be possible to follow even this simple strategy, but primer-dimer formation by the few exceptions may be controlled through appropriate choices of DNA polymerase and reaction conditions, as discussed below.

Sequence homology searches such as BLAST (http://www.ncbi.nlm.nih.gov/BLAST) can reveal if a primer sequence has multiple sites of partial complementarity (particularly at its 3'-end) within the gene of interest, pseudogenes, or stretches of homology elsewhere in the genome. Minimizing the number of such sites will minimize the number of secondary priming sites and thus increase the specificity and subsequent effectiveness of amplification of the desired fragments. Finally, each primer candidate should be evaluated to minimize the potential for the formation of stable secondary structures (hairpins) and complexes with other primers in the multiplex, such as through palindromic sequences or stretches of multiple-GC base pairs.

Several software packages are available to assist in the selection of primer sequences, including: Oligo (National Biosciences, Plymouth, MN), Rightprime (BioDisk Software, San Francisco, CA; www.biodisk.com), and Amplify (W. Engels, University of Wisconsin, Madison, WI). The latter two were specifically developed for the Macintosh. As of May 1999, on-line sites included:

http://www-genome.wi.mit.edu/cgi-bin/primer/primer3_www.cgi
http://www.hgmp.mrc.ac.uk/data/www/Menu/Menus/primer-design.html
http://www.williamstone.com/primers
http://life.anu.edu.au/molecular/software/gprime.htm
http://genome-www2.stanford.edu/cgi-bin/SGD/web-primer

5.2.1.4 Primer Label

One final note is specific to the detection format described here. Because the PCR product pool will be analyzed with a panel of immobilized oligonucleotide probes, a label for each product strand is introduced during PCR. This is perhaps most simply done using 5'-biotinylated primers. Such primers can be readily generated with a biotin phosphoramidite during primer synthesis.

5.2.2 Reaction Components

In addition to the primer sequences, appropriate selection of DNA polymerase, magnesium ion concentration, and primer concentrations are key to the success of a multiplex PCR system, as these factors influence the specificity and reliability of each amplification. Other components, however, can also affect amplification efficiencies. Cycling parameters are discussed in Section 5.2.3.

5.2.2.1 DNA Polymerase

Taq DNA polymerase, the recombinant thermostable DNA polymerase from *Thermus aquaticus* (Lawyer et al. 1989), is perhaps the most commonly used DNA polymerase for single-target PCR. A more recent advance that has proven particularly advantageous for multiplex PCR systems is the chemically modified version of AmpliTaq DNA Polymerase, known as AmpliTaq Gold (Perkin-Elmer, Foster City, CA, www.appliedbiosystems.com). This modified enzyme is inactive at room temperature and, as a result, transient primer-primer and primer-template complexes are not converted to products that can subsequently serve as template molecules. Heat restores AmpliTaq Gold's polymerase activity during PCR (Birch et al. 1996). By essentially eliminating primer-dimer formation and pre-PCR mispriming, AmpliTaq Gold may facilitate the optimization of primer sequences relative to a multiplex system utilizing the unmodified AmpliTaq DNA Polymerase.

Another option is AmpliTaq DNA Polymerase, Stoffel Fragment (Perkin-Elmer), which specifically lacks the 5′-to-3′ exonuclease activity intrinsic to the *Taq* DNA polymerase holoenzyme. Stoffel Fragment has greater thermal stability and optimal polymerase activity over a broader range of magnesium ion concentrations than the holoenzyme (Lawyer et al. 1993), which can be beneficial for multiplex PCR. Increased thermal stability of the polymerase is advantageous for systems with GC-bp-rich templates that require high denaturation temperatures, and flexibility in magnesium ion concentrations can simplify optimization of amplification conditions for an increasing number of fragments. As a consequence of its amino-terminal truncation, Stoffel Fragment will also extend by strand displacement through hairpin structures formed within template strands; in contrast, the 5′-to-3′ exonuclease activity of the holoenzyme is able to hydrolyze such structures (Abramson 1995), thus degrading template strands. Although less processive than the holoenzyme, the Stoffel Fragment has been particularly effective for multiplexes that must amplify fragments high in GC-bp content (Zangenberg et al. 1999).

Optimal enzyme concentrations are likely to increase as the number of fragments to be amplified increases, but initial testing can begin at 5 to 10 units per 100-μl PCR. As with single-target PCR, obtaining sufficiently high and comparable yields of all targets will depend on optimizing both enzyme concentration and extension times. Excess polymerase activity in early cycles can contribute to non-target product formation; insufficient activity in later cycles will compromise product yields. The time-release nature of AmpliTaq Gold DNA polymerase is another advantage of this enzyme (Birch et al. 1996; Kebelmann-Betzing et al. 1998).

5.2.2.2 Buffer and Monovalent Ion

The choice of DNA polymerase will generally determine the optimal buffer and monovalent ion to be used. The standard buffer for AmpliTaq DNA Polymerase is 10mM Tris-HCl (pH 8.3 at 25 °C) with 50mM KCl. The standard buffer for Stoffel Fragment is 10mM Tris-HCl with 10mM KCl. AmpliTaq Gold is more effectively activated in 15mM Tris-HCl (pH 8.0 at 25 °C) with 50mM KCl (Zangenberg et al. 1999).

Buffer additives that can facilitate the amplification of fragments high in GC-bp content include dimethyl sulfoxide (DMSO), glycerol (Pomp and Medrano 1991; Landre et al. 1995), and betaine (Weissensteiner and Lanchbury 1996), each of which lowers effective melting and strand separation temperatures. DMSO reduces the thermal stability of AmpliTaq DNA Polymerase (Landre et al. 1995), but can be more effective than glycerol as an additive for PCR. Chamberlain and Chamberlain (1994) reported greater routine success with multiplex PCR when using DMSO. Betaine stabilizes AT base pairs relative to GC base pairs (Rees et al. 1993), and may enhance the thermal stability of the DNA polymerase (Santoro et al. 1992).

5.2.2.3 Deoxynucleoside Triphosphates (dNTPs)

For single-target PCRs, dNTPs are generally used at 200 µM each (N = A, G, C, and T). Multiplex PCRs can utilize the same concentrations, although some groups report higher yields with 400 to 500µM each dNTP (Chamberlain and Chamberlain 1994; Henegariu et al. 1997). If uracil-N-glycosylase (UNG) is used to eliminate PCR product carryover contamination (Longo et al. 1990), then TTP should be replaced with two to three times as much dUTP (e.g., replace 200µM TTP with 400–600µM dUTP) to compensate for possible reductions in polymerase efficiency.

Base analogs such as 2'-deoxyinosine (dITP; Turner and Jenkins 1995) and 7-deaza-2'-deoxyguanosine (deaza-dGTP; McConlogue et al. 1988) can partially replace dGTP to relieve difficulties in amplifying fragments of high GC-bp content. These analogs destabilize GC base pairs and thus reduce the stability of secondary structures that template strands may form. A disadvantage of deaza-dGTP incorporation at high levels is the loss of fluorescence if ethidium bromide is used to stain gels (Weiss et al. 1994), but high levels may not be necessary if buffer additives or Stoffel Fragment are used in conjunction with a limited dGTP replacement.

5.2.2.4 Magnesium Chloride

Magnesium ions are necessary for polymerase function, and 1.5 mM $MgCl_2$ is generally used for PCR with 200 µM each dNTP. The $MgCl_2$ concentration will affect both reaction specificity and product yields and, therefore, should be optimized for each PCR system. Multiplex PCR systems often benefit from up to 10 mM $MgCl_2$ (Chamberlain and Chamberlain 1994; Henegariu et al. 1997; Zangenberg et al. 1999), although much lower concentrations should be chosen for single-target amplifications to evaluate the individual primer pairs prior to pooling them. If significant changes in total dNTP concentration are made, the $MgCl_2$ concentration may need to be adjusted accordingly, to account for the magnesium ions chelated by the dNTPs.

5.2.2.5 Primer Concentration

Optimal primer concentrations must be determined empirically for each multiplex system, since final product yields are affected by primer-primer and primer-template interactions, as well as relative amplification efficiencies of the multiplex targets. Initial concentrations can be estimated from single primer pair reactions at 0.1 to 0.2 µM each primer. For example, two primer pairs that result in equivalent product yields of similarly sized fragments will most likely be similar in concentration in the final multiplex, while primer pairs for longer targets will most likely be at higher concentrations to compensate for the generally lower yields of longer products. To obtain comparable yields of all products, final primer concentrations may vary significantly; in the CVD example described below, the final range for 14 primer pairs was 0.04–0.75 µM.

More recent work suggests that use of the common single-nucleotide at the 3′-end of the primers (Zangenberg et al. 1999) in combination with high (calculated) melting temperatures can yield sets of 10–40 primers that will work together at identical concentrations, thus eliminating the need for individual primer adjustments and a wide range of primer concentrations (R. Reynolds, unpubl.). Changes to reaction conditions that affect relative amplification efficiencies (e.g., annealing temperature, $MgCl_2$ concentration) may necessitate readjustment of some primer concentrations.

5.2.2.6 Template DNA

In general, 20 to 50 ng of genomic DNA per 100-µl multiplex PCR should be sufficient; this amount is certainly sufficient to evaluate primer candidates and initial primer pools. Multiplex PCR conditions can be optimized to enable reliable amplification from much less material than this, if necessary, but much greater amounts of template DNA should be avoided unless the sample is significantly degraded. Excess template (>100 ng) can contribute to non-target product formation, and can result in reduced product yields if an inhibitor is present. DNA preparation methods such as the salting-out method (Miller et al. 1988) or the Puregene DNA isolation kit (Gentra Systems, Minneapolis, MN, www.gentra.com) will yield microgram amounts of DNA from 5 or 10 ml of heparin-free blood, but it is best not to use more DNA than necessary for analysis. Crude DNA extractions, such as from buccal cells, can also be used for multiplex amplifications (Richards et al. 1993), although multiplex PCRs may become more sensitive to the quality and quantity of template DNA as the number of targets to be amplified increases.

5.2.3 Cycling Parameters

In conjunction with optimized reaction components, appropriate cycling parameters will result in consistent, specific amplification of all targets with comparable yields suitable for analysis. Primer annealing temperature may be the most critical parameter because of the impact on specificity, although extension times and the number of cycles of amplification also influence final product yields. Temperatures and times noted below are intended as guidelines for empirical optimization of cycling parameters.

5.2.3.1 Pre-PCR Enzyme Activation or Hot Start

AmpliTaq Gold requires a pre-PCR activation step, typically at 94 or 95°C. Maximum polymerase activity is restored within 15 min at 95°C in Tris-HCl (pH 8.3 at 25°C) buffer (Birch et al. 1996). Activation is more efficient at pH 8.0 (Zangenberg et al. 1999), such that this step can be reduced to 7–10 min. Excess enzyme activity during the early cycles of PCR can contribute to non-target product formation by strand synthesis at secondary priming sites. Consequently, several activation times should be compared to determine the optimum for reaction specificity with each multiplex system. If even greater specificity is required, due to a low-copy-number template, for example, then a shorter pre-PCR activation step can be combined with gradual enzyme activation during thermal cycling (Kebelmann-Betzing et al. 1998); additional cycles of amplification may be needed to increase final product yields.

A pre-PCR activation step is not required for either AmpliTaq DNA Polymerase or the Stoffel Fragment, although a 1- to 5-min step for denaturation of template strands is often used; the optimal time will depend on the model of thermal cycler used. A "hot start" procedure will minimize primer-dimer formation: withhold at least one essential reaction component (e.g., MgCl$_2$) until the temperature is sufficiently high to favor only specific primer-template annealing. Hot start protocols may employ a wax barrier (Chou et al. 1992) or a temperature-sensitive antibody to *Taq* DNA polymerase (Kellogg et al. 1994).

5.2.3.2 Denaturation

The denaturation step for each PCR cycle is typically between 15 and 60 s at 94 to 96 °C. The choice of temperature may depend on the base composition of the targeted fragments, as regions very high in GC-bp content may require a higher temperature for complete denaturation in the absence of buffer additives (Sect. 5.2.2.2). The denaturation time is primarily dependent on the type of thermal cycler used, and generally reflects the time needed for the reaction block to achieve thermal equilibrium.

5.2.3.3 Primer Annealing

The optimal annealing temperature will depend on the specific primer sequences, which, in turn, depend on the base composition of the target sequences. Higher temperatures result in greater reaction specificity by minimizing annealing at secondary priming sites, but may reduce some product yields from primers and targets with high AT-bp content. Annealing times are typically between 15 and 60 s, depending on the model of thermal cycler used. Longer times may improve product yields from a primer pair whose optimal annealing temperature is below the chosen multiplex annealing temperature.

5.2.3.4 Extension

Extension temperatures are usually 68 to 72 °C for optimal rates of strand synthesis by either *Taq* DNA polymerase or the Stoffel Fragment (Lawyer et al. 1993). The optimal extension time will depend on the size range of the fragments being amplified, as well as on the type of thermal cycler used (see Sect. 5.2.3.2). For product sizes below 1 kb, extension times of 30 to 60 s will generally suffice. As the number of targets being amplified increases, active enzyme molecules may become limiting at earlier cycle numbers, and product yields may be improved by lengthening the extension time during the later cycles. The "final extension" step of 5 to 10 min that is usually added after the last PCR cycle allows for completion of strand synthesis to yield fully double-stranded products.

5.2.3.5 Cycle Number

The number of cycles of amplification required will depend on the starting copy-number of the template DNA and the method of product analysis to be used. For gel analysis and the probe-based approach described in Section 5.3, amplification of 20 to 50 ng total genomic DNA (~6000–15,000 copies of nuclear DNA) for 32 to 35 cycles will be sufficient. Additional cycles may be required if AmpliTaq Gold DNA Polymerase is used in a time-release protocol (Kebelmann-Betzing et al. 1998), or if amplification efficiencies are reduced, as might be observed for long fragments.

5.3 Sequence-Specific Oligonucleotide Probe (SSOP) Methodology

As was demonstrated even before the advent of PCR, synthetic oligonucleotide probes whose sequences match known genetic variations of interest can be used to detect these variations with single-base discrimination (Wallace et al. 1979; Conner et al. 1983). When combined with PCR, SSOPs provide a highly sensitive and specific method to detect a wide variety of potentially disease-associated genetic changes, from single-base changes to larger insertions or deletions. The success of SSOP-based detection is based upon the choice of probe sequences and use of hybridization and stringent wash conditions to favor a perfect match between probe and target sequences and to discriminate against mismatched complexes. Key variables include the assay buffers, temperature, and time.

5.3.1 Detection Format

For a few variations and many samples, individual PCR products can be immobilized onto a solid support, such as a nylon membrane filter, using a vacuum spotting manifold to create a "dot blot" (Saiki et al. 1986; Saiki and Erlich 1998). These arrays are typically in a 96-well microtiter plate format. Each 96-spot array can then be queried with a series of SSOPs, or replicate filters can be prepared for each oligonucleotide probe. Hybridization and wash stringencies can be customized for each SSOP.

  As the number of sites of interest increases, however, it becomes preferable to have the probes immobilized onto the solid support, and to then query each probe panel with a different PCR product or product pool. This is also referred to as a "reverse dot blot" (Saiki et al. 1989; Saiki and Erlich 1998). One filter is used per sample, and multiple replicate filters can be prepared in advance. To further increase probe density and throughput for genotyping, the reverse dot blot can be expanded and modified to a line blot format, as in the CVD example presented here. Successful multiplexing of SSOPs depends on the choice of probe sequences and assay conditions to maximize the specificity of each probe for its intended target.

5.3.2 Probe Design

The SSOP approach is best suited to detect known, well-characterized genetic variation. For each site of interest, one probe should be designed specifically for each known variant; a biallelic site is therefore genotyped by a pair of probes. As with PCR primers, probe optimization is ultimately an empirical process, but there are guidelines for the selection of probe candidates.

5.3.2.1 Probe Length

Probes are typically between 15 and 25 nucleotides in length, depending on their base composition. Software packages available for primer design, such as Oligo, can be used to calculate melting temperatures for SSOPs; in principle, melting temperatures of 2–5 °C above the assay temperature (at 1 M Na$^+$) are best for single-base mismatch discrimination (Thein and Wallace 1986). Greater sequence discrimination may be possible with shorter probes than longer ones, because a single mismatch will be more destabilizing to a short probe-target complex, but optimal lengths will depend on the intended assay conditions. If probes are to be immobilized, stable secondary structures that a targeted PCR product strand is able to form may interfere with probe hybridization. A longer probe than would otherwise be recommended may be able to compete effectively; if not, the PCR product size may need to be reduced to minimize or eliminate the problematic sequence.

5.3.2.2 Probe Sequence

If each SSOP is specific for one genetic variant, the site of interest will create at least one mismatched base pair with all PCR strands except that specific variant; this is the source of probe discrimination among alleles. Mismatches will be most destabilizing if positioned in the center of the SSOP, and should be at least 3 bp from either end. If the sequence is particularly high in GC-bp content to one side of the site of interest, however, better discrimination may result if the probe site is offset towards the side that is less GC-rich, to minimize the number of these strong base pairs that will be formed by all alleles.

Depending on the particular base substitution at the site of interest, SSOPs may be from the same strand or from complementary strands. This is because mismatches differ in their relative instability. A G:T mismatch is relatively stable compared to an A:C mismatch (Thein and Wallace 1986), and is therefore less discriminating. If the base substitution of interest (A-to-G, G-to-A, T-to-C, or C-to-T) would result in a G:T mismatch between one allele and the probe designed to detect the other allele, then the complementary sequence should be used for probe design, to take advantage of the more destabilizing A:C mismatch. Similarly, a T:C mismatch is more discriminating than a G:A mismatch. These possibilities are most easily visualized by writing out the double-stranded sequence of the probe region and noting the base substitution of interest in each strand.

If probes are to be immobilized through a poly(T)-tail (Sect. 5.3.2.3), then the template sequence flanking the corresponding end of the probe should be examined to identify A-rich regions that could hybridize to the tail, effectively increasing the length and melting temperature of the probe. If such regions are identified, one solution is to add one or two non-complementary bases to that end of the intended probe sequence.

5.3.2.3 Probe Label

For the reverse dot or line blot formats discussed here, probe molecules are not labeled; instead, a 5′-biotin label is introduced into each strand generated by PCR. For immobilization of the SSOPs to a solid support, however, probes must be modified with a tether. One method of tethering an oligonucleotide probe is through a poly(T)-tail. Tails can either be added to the 5′-end (typically 100 T's) during probe synthesis, or to the 3′-end (typically 400 T's) using terminal deoxynucleotide transferase (Saiki et al. 1989). These tail lengths allow the SSOP to freely interact with PCR product strands during hybridization. For improved yields of 5′-poly(T)-tailed oligonucleotides, the capping reaction on the synthesizer can be disabled, since the precise number of thymidines is not critical (Saiki and Erlich 1998). Alternatively, probes may be conjugated at their 5′-ends to bovine serum albumin (BSA) using the methods described by Tung et al. (1991).

5.3.3 Probe Validation

All probes must be optimized for sensitivity and specificity of hybridization under the common assay conditions. In other words, each probe should correctly detect the allele for which it has been designed, and little if any signal should result from any other allele that may have been amplified. Probe candidates should be tested at several concentrations between 0.5 and 4 µM using genomic DNAs whose genotypes at the site of interest have been previously determined by means such as restriction endonuclease analysis or sequencing. Probe sequences may then need to be lengthened or shortened to achieve optimal signals. Once probe performance has been verified using single-target PCR products, then multiplex PCR product pools should be tested to optimize probe concentrations for adequate probe signals. Minor adjustments in primer concentrations may also be made to adjust product yields in conjunction with probe optimization. Furthermore, although signals from immobilized SSOPs are more qualitative than quantitative, heterozygous genotypes are most readily discerned if the probes for each site are adjusted to yield signals that are similar in strength.

In the absence of an existing genomic control DNA, two options are available. If the sequence variation is believed to be common, then probe signals obtained by screening 10 to 20 random samples may identify DNA samples that can then be sequenced to verify their genotype. Alternatively, a control template may be generated synthetically or by using mutagenic primers. For amplicons of fewer than 150 bp or so, a single-stranded template containing the rare sequence variation can be synthesized from phosphoramidites and then converted to the double-stranded form by PCR. For longer amplicons, "overlap extension" PCR (Higuchi et al. 1988; Ho et al. 1989) is more practical: Two overlapping and complementary mutagenic primers are used, independently paired with one of the primers for the full-length target to create a pair of double-stranded fragments (one upstream, one downstream) that overlap by 20 or more bases at the site of interest. In a third PCR, these overlapping fragments are annealed and then amplified using the outer primers to generate a full-length product bearing the variation of interest. Use of full-length controls is preferable to complementary oligonucleotides for probe optimization, because a short complement will not reveal whether the full-length PCR product could form stable secondary structures that would interfere with probe hybridization.

5.3.4 Allele Detection Conditions

As noted above, assay temperatures of 2–5 °C below the melting temperature of an SSOP are recommended for single-base mismatch discrimination. Assay temperatures are typically between 45 and 55 °C. The hybridization step, during which probe-target complexes are formed, is generally performed under permissive conditions to allow the denatured PCR product strands to freely associate with the immobilized SSOPs. The temperature and buffer are chosen to favor perfectly matched probe-target complexes, although the stringent wash is the most critical step for signal specificity. A starting hybridization condition is 20–30 min in a relatively high salt buffer such as 5X SSPE (0.9 M NaCl, 50 mM NaH_2PO_4, 5 mM EDTA, pH 7.4 at 25 °C) with 0.5% SDS. The stringent wash step might then be 10–15 min in a lower salt buffer such as 2.5X SSPE with 0.2% SDS. Higher temperatures, lower salt concentrations (e.g., 4X SSPE hybridization, 1.5X or 2X SSPE wash buffer), and longer wash times will each increase stringency and reduce nonspecific signal intensities; specific signals may also be affected. Several temperatures and/or salt concentrations should be compared to identify the assay condition compatible with the greatest number of probe candidates. All other probe sequences can then be modified as necessary to achieve maximal sensitivity and specificity under this common set of conditions.

Probe-target complexes can be detected colorimetrically. The CVD example below utilizes streptavidin to localize horseradish peroxidase to the now immobilized biotinylated PCR product. Substrates 3,3′,5,5′-tetramethylbenzidine and hydrogen peroxide then yield a blue precipitate indicative of successful probe-target hybridization. Colorimetric and chemiluminescent detection systems using streptavidin-alkaline phosphatase are also available (Bio-Rad, Hercules, CA, www.bio-rad.com; Amersham Pharmacia Biotech AB, Uppsala, Sweden, www.apbiotech.com).

5.4 Example: Candidate Markers of CVD Risk

The assay described here was designed to genotype up to 35 biallelic sites (Table 5.1) within 15 genes representing pathways implicated in the development and progression of atherosclerotic plaques: lipid metabolism, homocysteine metabolism, blood pressure regulation, thrombosis, and leukocyte adhesion. The reader is referred elsewhere (Cheng et al. 1998) for a detailed list of references for this marker set. These markers were chosen from literature reports, and include promoter sites, amino acid variations, and insertion/deletion variations believed to be in linkage disequilibrium with a functional genetic variation. The multiplex PCR and SSOP format of the assay restricted the choice of markers to those with sufficient sequence information to permit primer and probe design.

5.4.1 Multiplex PCR

5.4.1.1 Primers

The 35 sites were divided between two multiplex PCRs, because (1) an allele-specific approach was used to detect all possible combinations of codons 112 and 158 in the apoE gene (Cheng et al. 1998), (2) targets could be arbitrarily grouped as being either primarily related to lipid metabolism or not, and (3) this grouping corresponded to probe sets that were readily accommodated onto existing nylon membranes for maximal throughput of samples. Multiplex A consisted of 14 biotinylated primer pairs designed to amplify the e2 and e3 alleles of apoE, and targets within the apoB, apoCIII, CETP, LPL, and PON genes. Multiplex B consisted of 13 biotinylated primer pairs designed to amplify the e4 allele of apoE, and targets within the ACE, ATIIR$_1$, AGT, CBS, MTHFR, GPIIIa, fibrinogen, factor V, and ELAM genes. Primers for the CBS exon 8, factor V Leiden, and MTHFR targets have been published previously (Hu et al. 1993; Goyette et al. 1995; Ridker et al. 1995); all others were designed using the guidelines described in Section 5.2.1. Primer concentrations ranged from 0.04–0.75 μM to obtain generally comparable yields of all targets.

Final product sizes are listed in Table 5.1. These sizes allowed evaluation of relative product yields by agarose gel electrophoresis, as illustrated in Fig. 5.1. Longer amplicons were selected for the ACE gene to better balance amplification efficiencies between the insertion and deletion alleles. Two other initial amplicons were reduced by half to improve signal intensities from the corre-

Table 5.1. Panel of 35 biallelic sites implicated in CVD risk. Markers are grouped by gene within broad categories based on biological function. For each site, the wild-type or most frequent allele is listed to the left of the numerical nucleotide or codon position; the variant or less frequent allele is listed to the right. apo Apolipoprotein; CETP cholesteryl ester transfer protein; LPL lipoprotein lipase; PON paraoxonase; ACE angiotensin converting enzyme; AGT angiotensino-gen; ATIIR1 angiotensin II receptor type 1; CBS cystathionine beta-synthase; MTHFR methylene tetrahydrofolate reductase; GPIIIa glycoprotein IIIa; ELAM endothelial leukocyte adhesion molecule-1

Gene	Polymorphism(s)[a]	Product size (bp)
Lipid metabolism		
apoB	thr71ile	295
	arg3500gln mutation	272
apoCIII	T(–625)del	161
	C(–482)T, T(–455)C	163
	C1100T	195
	C3175G (SstI), T3206G	338
apoE	E3/E2 (cys112, arg158cys)	391
	E4 (112arg, arg158)	391
CETP	ile405val	141
	asp442gly	206
LPL	T(–93)G, T(–39)C	185
	asp9asn	220
	asn291ser	245
	ser447term	115
PON	gln192arg	95
Hypertension (renin-angiotensin system)		
ACE	Alu element ins/del	533/246
ATIIR₁	A1166C	185
AGT	met235thr	171
Homocysteine metabolism		
CBS	ile278thr and 68.bp del/ins, gly307ser	240/308
	Optional[b]: ala114val, arg125gln, glu131asp	104
MTHFR	C677T; optional[b]: C692T	196
Thrombosis		
Factor V	arg506gln (Leiden mutation)	223
Fibrinogen	β chain G(–455)A	290
GPIIIa	leu33pro	131
Leukocyte adhesion		
ELAM	G98T	118
	ser128arg	167
	leu554phe	160

[a]Codon changes designated by three-letter codes; nucleotide changes by single letters. Insertion (ins), deletion (del)
[b]Secondary probe strip because of the likely infrequency of these variations

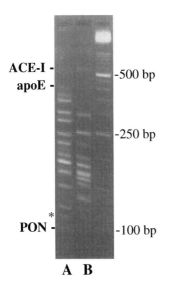

Fig. 5.1. Agarose gel image illustrating multiplex PCR product pools for CVD markers. Product sizes are given in Table 5.1; the longest and shortest fragments amplified by each multiplex are noted. The mobility standard is molecular weight marker XIII. Not all of the PCR products are clearly visible, and one low molecular weight non-target band is visible (marked with an *asterisk*)

sponding SSOPs; the longer initial PCR products may have formed stable secondary structures that inhibited hybridization with the immobilized probes. In addition, the apoCIII promoter fragment was most effectively amplified if divided into two separate amplicons of 161 and 163 bp. A full-length product encompassing this apoCIII promoter region can also be seen in Fig. 5.1.

5.4.1.2 PCR Conditions

As others have reported (Houlston et al. 1989; Hixson and Vernier 1990), amplification of the apoE region, which is relatively high in GC-bp content, was most efficient in the presence of DMSO. The final concentration of DMSO was selected to enable reliable amplification of the apoE alleles with minimal adverse impact on the yields of products that were high in AT-bp content.

Approximately 50 ng of total genomic DNA was used for each assay, 25 ng for each multiplex A and multiplex B reaction. In addition to the primer pools, each 50-µl reaction contained 20 mM Tris-HCl (0.2 M stock, pH 8.3 at 250 °C), 50 mM KCl, 8.5% DMSO (v/v), 0.1 mM dATP, 0.1 mM dCTP, 0.07 mM dGTP, 0.03 mM deaza-dGTP, 0.2 mM dUTP, 1.7 mM $MgCl_2$, and 7 units of AmpliTaq Gold. Deaza-dGTP (Roche Molecular Biochemicals, Indianapolis, IN, www.biochem.boehringer-mannheim.com) was incorporated to facilitate amplification of regions high in GC-bp content. DeoxyUTP was included for compatibility with the use of uracil *N*-glycosylase to eliminate PCR product contamination (Longo et al. 1990). For higher volume genotyping, these PCRs can be performed in 96-well Thermowell polypropylene plates with sealing mats (Corning Costar, Cambridge, MA, www.corningcostar.com).

Samples were amplified in a Perkin-Elmer GeneAmp PCR System 9600 using a 2.4-h thermal cycling profile: an initial hold of 94 °C for 12.5 min; then 33 cycles of 96 °C for 15 s, 60 °C for 1 min, and 72 °C for 1.25 min; and a final extension step of 68 °C for 5 min.

To evaluate relative product yields during assay development, 3- to 5-μl aliquots of each multiplex PCR were analyzed on horizontal gels made from 3% NuSieve and 1% SeaKem GTG agarose (FMC BioProducts, Rockland, ME, www.bioproducts.com) in 89 mM Tris-borate with 1 mM EDTA and ethidium bromide. Molecular weight marker XIII (Roche Molecular Biochemicals) was used as a mobility standard. As Fig. 5.1 illustrates, not all of the longer product bands are clearly visible; however, the corresponding markers could still be detected by the immobilized probes.

5.4.2 Immobilized SSOPs

Each biallelic site was represented by two probes, one for each variant. To confirm successful amplification of the two longest fragments, probes were also designed for invariant regions of the apoE and ACE amplicons. Following the guidelines described in Sections 5.3.2 and 5.3.3, the sequences and concentrations of the final 70 probes were chosen to achieve signal balance between alleles at each site, and for generally comparable intensities among all of the loci.

The addition of 3′-poly(T) tails to probes is a simple, overnight procedure using terminal deoxyribonucleotidyl transferase (TdT): In a 100-μl reaction, 200 pmol SSOP is combined with 80 nmol TTP in 1X TdT buffer (100 mM K-cacodylate, 25 mM Tris-HCl, pH 7.6, 1 mM $CoCl_2$, 0.2 mM dithiothreitol) with ~20 units TdT (Amersham Pharmacia Biotech AB). After overnight incubation (~15 h) at 37 °C, the reaction is stopped by the addition of 2 μl of 0.5 M EDTA, pH 8. T-tailed probes are stored at –20 °C. This ratio of probe to TTP molecules will add a tail of approximately 400 bases; tail length can be changed by adjusting this ratio.

Probes can be spotted onto nylon filters (e.g., 0.45 μM Biodyne B, Pall, Glen Cove, NY, www.pall.com) in solutions of 10 mM Tris-HCl (pH 8.0 at 25 °C) with 0.1 mM EDTA. Typical probe solutions are between 0.5 and 4.0 μM; to assist in visualizing the spots, an inert dye (e.g., 0.002% Orange II) can be included. Dot and slot blot vacuum manifolds are available from Bio-Rad or Life Technologies (Gaithersburg, MD, www.lifetech.com). As Fig. 5.2 illustrates, offsetting the membrane permits immobilization of additional probes using a standard 96-well dot blot vacuum manifold. Care must be taken to avoid air bubbles and to ensure even deposition of probe solution throughout the dot or slot. Deposited poly(T)-tailed probes should be irradiated with 254-nm light to 120 mJ/cm^2 while the nylon membrane is still damp to crosslink the thymine bases to the primary amines of the support (Saiki and Erlich 1998). Automated systems are also available (e.g., Robbins Scientific, Sunnyvale, CA,

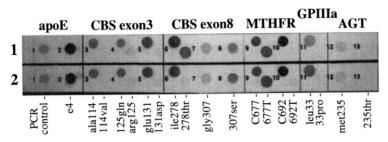

Fig. 5.2. Two sample strips with 3′-poly(T)-tailed probes in the reverse dot blot format. These strips are specific for the apoE, CBS, MTHFR, GPIIIa, and AGT alleles noted, all amplified by multiplex B primers. These two samples differ in their genotype at codon 278 of the CBS gene. The weak signals from CBS gly307 and AGT met235 probes, and the false positive signal from the CBS 307ser probe, indicate that this probe set has not been fully optimized. For the final assay, the gly307 probe was lengthened, the 307ser probe was shortened, and the AGT amplicon was reduced by half

www.robsci.com; BioDot, Irvine, CA, www.biodot.com; and IVEK, North Springfield, VT).

To enable larger-scale production of probe strips, the 70 probes for these 35 CVD markers were conjugated at their 5′-ends to BSA by methods similar to those used by Tung et al. (1991), then applied in a linear array to sheets of nylon membrane using an IVEK linear striper and Multispense200 controller (IVEK, N. Springfield, VT). As shown in Fig. 5.3, the probes on "probe strip A" corresponded to the targets amplified by multiplex A; "probe strips B and B2" corresponded to the targets amplified by multiplex B.

5.4.3 Allele Detection

Detection reagents developed for HLA genotyping using the same format described here are available from Dynal (Oslo, Norway, www.dynal.no) and were used to generate the data shown in Fig. 5.3; alternative formulations are given in the protocol below. Actual volumes of solution needed for each well will vary depending on the actual typing tray used; the solution should adequately cover the entire probe strip during the assay procedure, but must not splash into adjacent wells during agitation.

The final temperature of 52 °C used for detection of the amplified CVD alleles was necessary to improve specificity at the apoCIII (–625) site. This sequence variation is the presence (more frequent allele) or absence (less frequent allele) of an A:T base pair between a G:C base pair doublet and quartet within a region generally high in GC-bp content. Improved discrimination between apoCIII (–625) alleles was also observed at assay temperatures above 52 °C, but these higher temperatures adversely affected the signal intensities from probes for other markers (data not shown).

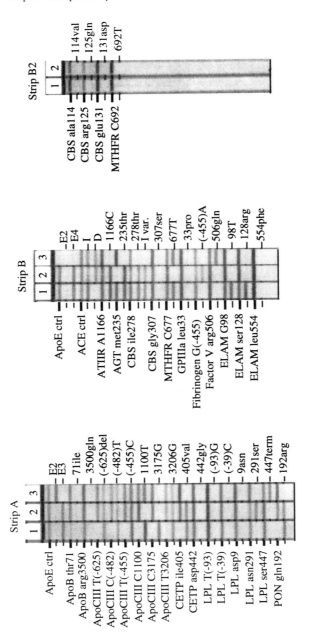

Fig. 5.3. Representative probe strips in the line format. Three examples each of probe strips A and B are shown; only two examples of probe strip B2 are shown. On each vertical strip, a *horizontal line* (blue in actual color) indicates detection of the corresponding allele amplified from the original sample. The template guides to the left and right of each image identify the specific allele detected at each probe position. With the exception of apoE3, ACE-I and ACE-D, the less prevalent genetic variant for each marker is listed on the right. The probe strip A image illustrates detection of each genotype for apoCIII nucleotide 1100, CETP codon 405, and PON codon 192. The probe strip B image illustrates detection of the three possible AGT codon 235 genotypes. Differences in the relative efficiencies of amplification and probe-hybridization contribute to the variation in signal intensity among the loci. In actual size, each strip is ~8 cm long

Allele detection is performed using a water bath rotating at 50–60 rpm (e.g., Hot Shaker Plus from Bellco, Vineland, NJ, www.bellcoglass.com). Probe strips are pre-washed at 52 °C for 10 min to remove unbound probe in 2X SSPE (0.36 M NaCl, 0.02 M Na$_2$HPO$_4$, 2 mM EDTA, adjusted to pH 7.4 with NaOH), 0.5% SDS. Twenty-μl aliquots of the biotinylated PCR product pools from multiplex A and B reactions are mixed with equal volumes of a denaturation solution (0.4 N NaOH, 20 mM EDTA), then added to typing tray (Perkin-Elmer) wells containing 3 ml of hybridization buffer (4X SSPE, 0.5% SDS) and a correspondingly labeled probe strip A or probe strip B. Probe strip B2 is placed alongside strip B. After 20 min at 52 °C, the hybridization solution is replaced with fresh buffer containing 10 μl of streptavidin-horseradish peroxidase conjugate (SA-HRP; Perkin-Elmer) and the strips are returned to the water bath for 5 min. This enzyme conjugate solution is then replaced with the stringent wash buffer (2X SSPE, 0.5% SDS), and the strips are returned to 52 °C for 12 min. The washed strips are equilibrated in 50 mM Na-citrate (pH 5) for 5–10 min at room temperature on a rotating (50–60 rpm) platform (e.g., Gyrotory shaker Model G2 from New Brunswick Scientific, Edison, NJ, www.nbsc.com).

Color development reagent can be prepared from 100 mM citrate solution, 2 mg/ml 3,3′,5,5′-tetramethylbenzidine (TMB) in ethanol (Perkin-Elmer), and 3% H$_2$O$_2$ in water, mixed in a 20 ml/1 ml/20 μl ratio (Saiki and Erlich 1998). Sigma-Aldrich (St. Louis, MO, www.sigma-aldrich.com) sells a TMB liquid substrate system containing both TMB and H$_2$O$_2$. The citrate equilibration buffer is replaced with an equal volume of the color-development reagent and strips are gently agitated for 8–10 min at room temperature. With longer times, strips will acquire a blue background. Developed strips must be rinsed thoroughly with water to stop the colorimetric reaction.

Developed strips should be aligned on a flat surface next to a guide identifying the allele detected by each probe line, and photographed using type 559 or type 55 film from Polaroid (Cambridge, MA, www.polaroid.com). Alternatively, strips can be scanned and the images stored electronically. Genotype interpretations should be made independently by two individuals. An SLT ProfiBlot IIT (Tecan US, Research Triangle Park, NC, www.tecan.com) can also be employed to automate the hybridization, stringent wash, and color-development steps.

Immobilized probe filters should not be reused because of the tendency for PCR product strands to adhere to the nylon; once dried, removal can be difficult. If necessary, just-used filters can be decolorized by gentle agitation for up to an hour in 1–2% Na-sulfite solution at room temperature. Decolorized filters must be rinsed with water, then gently boiled twice for 5–10 min each in 10 mM Tris-HCl (pH 8.3), 0.1 mM EDTA, 0.1% SDS, with a buffer change between these steps. After a final rinse in water, filters can be dried. A "stripped" filter can be redeveloped by equilibrating it in citrate solution, then adding color-development solution, to confirm the removal of biotinylated PCR product; completely stripped filters can be decolorized once more, then rinsed carefully before re-use.

5.5 Summary

Well-defined phenotypes, both quantitative intermediate phenotypes as well as clinical endpoints, and multilocus genetic data will be essential for studies of disease predisposition factors. With multiplex PCR and immobilized probes, cohorts can be rapidly queried at multiple biallelic sites, providing the epidemiological data needed to evaluate the association of these markers with disease or therapeutic response. One limitation of this approach is that sequence-specific probes will generally not identify new mutations or polymorphisms; only those new sequence variations resulting in unusually weak signal intensities or apparently null genotypes would be detected. Furthermore, this format is not suited for variable number tandem repeat polymorphisms. Higher density probe arrays (Chap. 7) are more appropriate for the detection of specific mutations in genes such as the LDL receptor gene, for which over 600 mutations, including large deletions, have been reported (http://www.ucl.ac.uk/fh). As a research tool, however, the reverse blot format offers significant flexibility for customizing a panel of markers. Informative markers identified by the research community may someday form the basis of clinical panels for diagnostic applications.

Acknowledgements. The CVD assay described here was developed with the excellent technical support of Michael Grow and the Oligo Synthesis and Sequencing Groups at Roche Molecular Systems. I would also like to thank H. Erlich, R. Reynolds, R. Saiki, and G. Zangenberg for their generous support and technical advice, and R. Saiki for his comments on this chapter as well.

References

Abramson RD (1995) Thermostable DNA polymerases. In: Innis MA, Gelfand DH, Sninsky JJ (eds) PCR strategies. Academic Press, San Diego, pp 39–57

Birch DE, Kolmodin L, Wong J, Zangenberg GA, Zoccoli MA (1996) Simplified hot start PCR. Nature 381:445–446

Chamberlain JS, Chamberlain JR (1994) Optimization of multiplex PCRs. In: Mullis KB, Ferré F, Gibbs RA (eds) The polymerase chain reaction. Birkhäuser, Boston, pp 38–46

Chamberlain JS, Gibbs RA, Ranier JE, Nguyen PN, Caskey CT (1988) Deletion screening of the Duchenne muscular dystrophy locus via multiplex DNA amplification. Nucleic Acids Res 16:11141–11156

Cheng S, Pallaud C, Grow MA, Scharf SJ, Erlich HA, Klitz W, Pullinger CR, Malloy MJ, Kane JP, Siest G, Visvikis S (1998) A multilocus genotyping assay for cardiovascular disease. Clin Chem Lab Med 36:561–566

Chou Q, Russell M, Birch DE, Raymond J, Bloch W (1992) Prevention of pre-PCR mis-priming and primer dimerization improves low-copy-number amplifications. Nucleic Acids Res 20: 1717–1723

Conner BJ, Reyes AA, Morin C, Itakura K, Teplitz RL, Wallace RB (1983) Detection of sickle cell β^s-globin allele by hybridization with synthetic oligonucleotides. Proc Natl Acad Sci USA 80: 278–282

Goyette P, Frosst P, Rosenblatt DS, Rozen R (1995) Seven novel mutations in the methylenete-trahydrofolate reductase gene and genotype/phenotype correlations in severe methylenete-trahydrofolate reductase deficiency. Am J Hum Genet 56:1052–1059

Henegariu O, Heerema NA, Dlouhy SR, Vance GH, Vogt PH (1997) Multiplex PCR: critical para-meters and step-by-step protocol. Biotechniques 23:504–511

Higuchi R, Krummel B, Saiki RK (1988) A general method of *in vitro* preparation and specific mutagenesis of DNA fragments: study of protein and DNA interactions. Nucleic Acids Res 16:7351–7367

Hixson JE, Vernier DT (1990) Restriction isotyping of human apolipoprotein E by gene amplifi-cation and cleavage with *Hha*I. J Lipid Res 31:545–548

Ho SN, Hunt HD, Horton RM, Pullen JK, Pease LR (1989) Site-directed mutagenesis by overlap extension using the polymerase chain reaction. Gene 77:51–59

Hoeg JM (1997) Evaluating coronary heart disease risk. JAMA 277:1387–1390

Houlston RS, Snowden C, Green F, Alberti KGMM, Humphries SE (1989) Apolipoprotein (apo) E genotypes by polymerase chain reaction and allele-specific oligonucleotide probes: no detectable linkage disequilibrium between apo E and apo CII. Hum Genet 83:364–368

Hu FL, Gu Z, Kozich V, Kraus JP, Ramesh V, Shih VE (1993) Molecular basis of cystathionine β-synthase deficiency in pyridoxine responsive and nonresponsive homocystinuria. Hum Mol Genet 2:1857–1860

Kebelmann-Betzing C, Seeger K, Dragon S, Schmitt G, Möricke A, Schild TA, Henze G, Beyermann B (1998) Advantages of a new *Taq* DNA polymerase in multiplex PCR and time-release PCR. Biotechniques 24:154–158

Kellogg DE, Rybalkin I, Chen S, Mukhamedova N, Vlasik T, Siebert PD, Chenchik A (1994) Taq-Start antibody: "hot start" PCR facilitated by a neutralizing monoclonal antibody directed against *Taq* DNA polymerase. Biotechniques 16:1134–1137

Landre PA, Gelfand DH, Watson RM (1995) The use of cosolvents to enhance amplification by the polymerase chain reaction. In: Innis MA, Gelfand DH, Sninsky JJ (eds) PCR strategies. Aca-demic Press, San Diego, pp 3–16

Lawyer FC, Stoffel S, Saiki RK, Myambo K, Drummond R, Gelfand DH (1989) Isolation, charac-terization, and expression in *Escherichia coli* of the DNA polymerase gene from *Thermus aquaticus.* J Biol Chem 264:6427–6437

Lawyer FC, Stoffel S, Saiki RK, Chang S-Y, Landre PA, Abramson RD, Gelfand DH (1993) High-level expression, purification, and enzymatic characterization of full-length *Thermus aquati-cus* DNA polymerase and a truncated form deficient in 5′ to 3′ exonuclease activity. PCR Methods Appl 2:275–287

Li H, Cui X, Arnheim N (1990) Direct electrophoretic detection of the allelic state of single DNA molecules in human sperm by using the polymerase chain reaction. Proc Natl Acad Sci USA 87:4580–4584

Longo MC, Berninger MS, Hartley JL (1990) Use of uracil DNA glycosylase to control carry-over contamination in polymerase chain reactions. Gene 93:125–128

Marenberg ME, Risch N, Berkman LF, Floderus B, de Faire U (1994) Genetic susceptibility to death from coronary heart disease in a study of twins. N Engl J Med 330:1041–1046

Maron BJ, Moller JH, Seidman CE, Vincent GM, Dietz HC, Moss AJ, Towbin JA, Sondheimer HM, Pyeritz RE, McGee G, Epstein AE (1998) Impact of laboratory molecular diagnosis on contemporary diagnostic criteria for genetically transmitted cardiovascular diseases: hypertrophic cardiomyopathy, long-QT syndrome, and Marfan syndrome. Circulation 98:1460–1471

McConlogue L, Brow MAD, Innis MA (1988) Structure-independent DNA amplification by PCR using 7-deaza-2′-deoxyguanosine. Nucleic Acids Res 16:9869

Miller SA, Dykes DD, Polesky HF (1988) A simple salting out procedure for extracting DNA from human nucleated cells. Nucleic Acids Res 16:1215

Mullis KB, Faloona FA (1987) Specific synthesis of DNA in vitro via a polymerase-catalyzed chain reaction. Methods Enzymol 155:335–350

Pasternak RC, Grundy SM, Levy D, Thompson PD (1996) 27th Bethesda Conference: matching the intensity of risk factor management with the hazard for coronary disease events. Task Force 3. Spectrum of risk factors for coronary heart disease. J Am Coll Cardiol 27:978–990

Pomp D, Medrano JF (1991) Organic solvents as facilitators of polymerase chain reaction. Biotechniques 10:58–59

Rees WA, Yager TD, Korte J, von Hippel PH (1993) Betaine can eliminate the base pair composition dependence of DNA melting. Biochemistry 32:137–144

Richards B, Skoletsky J, Shuber AP, Balfour R, Stern RC, Dorkin HL, Parad RB, Witt D, Klinger KW (1993) Multiplex PCR amplification from the CFTR gene using DNA prepared from buccal brushes/swabs. Hum Mol Genet 2:159–163

Ridker PM, Hennekens CH, Lindpaintner K, Stampfer MJ, Eisenberg PR, Miletich JP (1995) Mutation in the gene coding for coagulation factor V and the risk of myocardial infarction, stroke, and venous thrombosis in apparently healthy men. N Engl J Med 332:912–917

Rychlik W, Spencer WJ, Rhoads RE (1990) Optimization of the annealing temperature for DNA amplification in vitro. Nucleic Acids Res 18:6409–6412

Saiki RK, Erlich HA (1998) Detection of mutations by hybridization with sequence-specific oligonucleotide probes. In: Cotton RGH, Edkins E, Forrest S (eds) Mutation detection: a practical approach. Oxford University Press, Oxford, pp 113–129

Saiki RK, Bugawan TL, Horn GT, Mullis KB, Erlich HA (1986) Analysis of enzymatically amplified beta-globin and HLA-DQ alpha DNA with allele-specific oligonucleotide probes. Nature 324:163–166

Saiki RK, Gelfand DH, Stoffel S, Scharf SJ, Higuchi R, Horn GT, Mullis KB, Erlich HA (1988) Primer-directed enzymatic amplification of DNA with a thermostable DNA polymerase. Science 239:487–491

Saiki RK, Walsh PS, Levenson CH, Erlich HA (1989) Genetic analysis of amplified DNA with immobilized sequence-specific oligonucleotide probes. Proc Natl Acad Sci USA 86:6230–6234

Santoro MM, Liu Y, Khan SMA, Hou L-X, Bolen DW (1992) Increased thermal stability of proteins in the presence of naturally occurring osmolytes. Biochemistry 31:5278–5283

Schafer AJ, Hawkings JR (1998) DNA variation and the future of human genetics. Nat Biotechnol 16:33–39

ten Kate LP, Boman H, Daiger SP, Motulsky AG (1982) Familial aggregation of coronary heart disease and its relation to known genetic risk factors. Am J Cardiol 50:945–953

Thein SL, Wallace RB (1986) The use of synthetic oligonucleotides as specific hybridization probes in the diagnosis of genetic disorders. In: Davies KE (ed) Human genetic diseases. IRL Press, Oxford, pp 33–50

Tung CH, Rudolph MJ, Stein S (1991) Preparation of oligonucleotide-peptide conjugates. Bioconjug Chem 2:464–465

Turner SL, Jenkins FJ (1995) Use of deoxyinosine in PCR to improve amplification of GC-rich DNA. Biotechniques 19:48–52

Wallace RB, Shaffer J, Murphy RF, Bonner J, Hirose T, Itakura K (1979) Hybridization of synthetic oligodeoxyribonucleotides to $\Phi\chi 174$ DNA: the effect of single base pair mismatch. Nucleic Acids Res 6:3543–3557

Walsh PS, Erlich HA, Higuchi R (1992) Preferential PCR amplification of alleles: mechanisms and solutions. PCR Methods Appl 1:241–250

Weiss J, Zucht H-D, Forssmann W-G (1994) Amplification of gene fragments with very high G/C content: c^7dGTP and the problem of visualizing the amplification products. PCR Methods Appl 4:124–125

Weissensteiner T, Lanchbury JS (1996) Strategy for controlling preferential amplification and avoiding false negatives in PCR typing. Biotechniques 21:1102–1108

Zangenberg G, Saiki RK, Reynolds R (1999) Multiplex PCR: optimization guidelines. In: Innis MA, Gelfand DH, Sninsky JJ (eds) PCR applications: protocols for functional genomics. Academic Press, San Diego, pp 73–94

6 The Special Case of HLA Genes: Detection and Resolution of Multiple Polymorphic Sites in a Single Gene

W. Martin Howell and Katherine L. Poole

6.1 Introduction

The major histocompatibility complex (MHC) has been the most intensively studied region of the human genome during the past three decades (Bodmer 1995) and it has also been widely studied in other mammalian species, including primates (Trowsdale 1995). These studies suggest that the MHC is probably the area of the mammalian genome that is most densely populated with functional genes. In humans, the MHC is principally known as the region that includes genes encoding the human leukocyte antigens (HLA), which play a crucial role in regulating the immune response in health and disease. These HLA genes are the most polymorphic found in humans and, since most polymorphisms encode functional variants, this polymorphism results in inter-individual variation in the immune response. As a result of this, HLA polymorphism has long been recognised as a critical factor in determining allograft acceptance and rejection mechanisms in both renal and bone marrow transplantation (Opelz et al. 1993; Madrigal et al. 1997b), while particular HLA genotypes are associated with a large number of benign and malignant immunologically mediated diseases (Thorsby 1997; Bateman and Howell 1999). Although the MHC also encodes a number of other genes of immunological significance (e.g. the tumour necrosis factor genes), this chapter will focus solely upon the so-called "classical" HLA class I and class II antigens. In particular, the extent, characteristics and distribution of polymorphism within the HLA genes and the need for methods for determining HLA genotypes in genetic epidemiological studies of a large number of diseases and in clinical transplantation will be considered. An overview of some of the wide range of PCR-based methods now available for the detection and resolution of HLA polymorphisms will be given, with particular emphasis on the demands of throughput and allelic resolution required, including the problem of distinguishing multiple heterozygous sites in a single gene. Finally, a practical strategy for HLA typing in retrospective disease association studies using archival biopsy tissues as a source of DNA for analysis will be presented. Such an approach is applicable to all other genetic systems in which multiple polymorphic sites occur within a single gene. HLA is a special case only in the

Principles and Practice
Molecular Genetic Epidemiology – A Laboratory Perspective
Ian N.M. Day (Ed.)
© Springer-Verlag Berlin Heidelberg 2002

exceptional degree of polymorphism that occurs in each of several closely linked genes. The HLA system may therefore be regarded as the most demanding testing ground for any molecular genotyping approach.

6.2 The HLA System

The "classical" HLA molecules are encoded by two highly polymorphic gene families located within a 3600-kb region of the MHC on chromosome 6 (6p21.3). The resulting HLA molecules are membrane-bound glycoproteins that bind processed antigenic peptides and bind them to T cells. The HLA class I A, B and C molecules are each composed of an MHC-encoded heavy chain (MW 45 kDa), non-covalently associated with a non-polymorphic polypeptide, β_2-microglobulin (MW 12 kDa) encoded on chromosome 15. There are now known to be 124 HLA-A alleles, 258 HLA-B alleles and 74 HLA-C alleles, including silent substitutions and null alleles, although most of these alleles encode different functional variants (Bodmer et al. 1999). Many alleles are found at a finite frequency in most populations and heterozygosity is high. HLA class I antigens are expressed on all nucleated cells (except foetal trophoblast) and platelets and their function is to present intracellularly processed peptides of largely endogenous (viral) origin to CD8+ T cells, which in the main are cytotoxic. The bound peptides are highly circumscribed in length, usually 8–9 amino acids, and are held in a peptide-binding groove, which X-ray crystallography has shown to have an allele-specific conformation (Stern and Wiley 1994). The distinctive polymorphic residues, which are encoded by the different alleles of a particular HLA class I gene, are found almost exclusively within this peptide-binding groove (Parham 1992). In addition, HLA class I epitopes also interact with both killer-inhibitory and killer-activating receptors expressed on natural killer and some T cells (Lanier 1998).

In contrast to class I molecules, HLA class II molecules, comprising three main subclasses – DR, DQ and DP – are found on a more restricted range of cell types, including B cells, activated T cells, the monocyte/macrophage lineage and are also interferon γ inducible. An expressed HLA class II molecule consists of an α-chain (MW 31–34 kDa) encoded by an A gene, non-covalently associated with a β-chain (MW 26–29 kDa), encoded by a B gene. The DQ and DP subregions both consist of a single expressed A gene (DQA1 and DPA1) and a single expressed B gene (DQB1 and DPB1). The DR subregion also consists of a single A gene (DRA), but may consist of one or more expressed B genes, depending on the DR haplotype. Each chromosome 6 encodes the expressed DRB1 gene, while one (but not more than one) of the DRB3, DRB4 or DRB5 genes may also be present (Trowsdale 1995). Both A and B genes may be polymorphic, but most of the polymorphism resides in the B genes (HLA-DRA is non-polymorphic in the region of the gene encoding the antigen-binding groove of the molecule). Currently, there are 2 DRA, 264 DRB,

20 DQA1, 39 DQB1, 15 DPA1 and 85 DPB1 alleles known, the vast majority of which encode expressed and most probably functional variations (Bodmer et al. 1999). Both α- and β-chains combine to form a peptide-binding groove, shown by X-ray crystallography to be very similar to the class I groove (Brown et al. 1993). However, class II molecules present peptides, mostly of exogenous origin, to CD4+ T cells, of largely "helper" phenotype. These bound peptides are generally longer and more variable in length than peptides bound to class I molecules (i.e. 14–25 amino acids), due to the more "open" ends of the peptide-binding groove.

The application of molecular techniques of gene mapping, cloning and sequencing has revealed a number of additional HLA and non-HLA genes in the class I and class II regions, some of which are shown in Fig. 6.1. Some of these genes, e.g. HLA-E and HLA-G, play specialised roles in antigen presentation (Braud et al. 1999), while others, including the TAP, LMP and HLA-DM and DO genes, play critical roles in the pathways of antigen processing for presentation by the "classical" HLA class I and class II molecules (Brodsky et al. 1996; Vogt et al. 1997; van Ham et al. 1997). These loci also exhibit a degree of polymorphism, but this is limited compared with that of the "classical" class I and II genes themselves. As yet, a possible role for polymorphism of these "non-classical" HLA genes in modulating disease pathogenesis remains unknown, but is under active investigation in many laboratories worldwide (Singal and

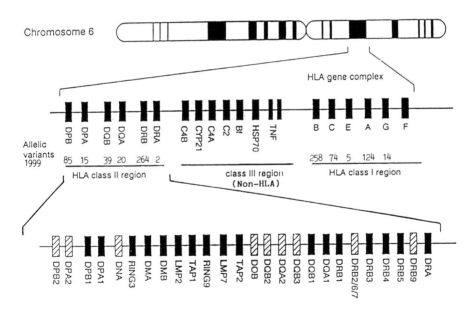

Fig. 6.1. The HLA gene complex on chromosome 6p21.3. Loci known to be expressed are marked with *filled squares*. Numbers of alleles are derived from Bodmer et al. (1999). (Adapted from Thorsby and Ronningen 1992)

Ye 1998; Tuokko et al. 1998). While characterisation and detection of these polymorphisms is outside the scope of this chapter, the same methods which can be used to detect polymorphisms of the "classical" HLA class I and class II genes may also be used for the detection of polymorphism among the 'non-classical MHC genes, such as TAP, LMP and HLA-DM (Powis and Teisserenc 1997).

6.3 Requirement for HLA Genotyping: Historical Perspectives

6.3.1 HLA, Disease and Clinical Transplantation

Due to the critical role of HLA molecules in regulating the immune response and the exuberant polymorphism of the HLA system, it is unsurprising that particular HLA alleles are associated with a wide range of immunologically mediated diseases, including autoimmune, infectious, allergic and neoplastic diseases. The first HLA and disease association – involving Hodgkin's disease – was described in 1967 (Amiel 1967) and since then a growing number of diseases have been shown to occur in individuals expressing particular HLA polymorphisms. Some of these associations – such as that between the HLA-DQA1*0501 and DQB1*0201 alleles and coeliac disease – are very strong, but the majority are much weaker – such as those involving particular HLA-DRB1*04 alleles and rheumatoid arthritis and DQA1/DQB1-encoded epitopes and insulin-dependent diabetes (reviewed by Thorsby 1997). The same applies to the documented associations involving infectious (Lalvani and Hill 1998), allergic (reviewed by Howell and Holgate 1996) and neoplastic diseases (reviewed by Bateman and Howell 1999). Such investigations are ongoing and one current aim is to move from case-control to family-based studies where segregating alleles and haplotypes can be investigated using the transmission disequilibrium test (Spielman et al. 1993) and to determine the genetic contribution of non-HLA genes to HLA-associated diseases. In any event, genetic epidemiological studies of any of these conditions must include an appropriate level of HLA genotyping. In addition, a particular interest of our laboratory – and many others – is to determine the role HLA polymorphism may play in determining disease outcome as well as initiation. For such studies, DNA-based analysis of HLA polymorphism using DNA derived from archival tissues is an attractive option, since – unlike prospective studies – clinical and pathological follow-up data of many years duration may already be available (Bateman et al. 1998).

Very early in the history of clinical renal transplantation it became clear that patient-donor HLA mismatching had a deleterious effect on allograft survival, due to recognition of non-self HLA molecules by the recipient's immune system (reviewed by Dyer and Claas 1997). The reverse situation arises in allogeneic bone marrow transplantation, whereby T cells in the donor bone marrow (transplanted into an immunocompromised host) recognise the HLA molecules expressed on recipient cells as foreign, so initiating graft-versus-host disease (GVHD), which can be fatal. Therefore accurate HLA typing prior to allogeneic bone marrow transplantation is crucial to ensure the best possible match of donor and recipient, thereby decreasing the chance of rejection and GVHD (Madrigal et al. 1997a).

Accordingly, the needs of the clinical laboratory, where accurate HLA geno-typing at an appropriate resolution is a requirement for both solid organ and especially bone marrow transplant matching programmes, have provided much impetus not only for elucidation of the genetics and polymorphism of the human MHC, but have also generated an ongoing revolution in techniques used for HLA genotyping, which consequently serve both clinical and research needs (Howell and Navarrete 1996; Madrigal et al. 1997a).

6.3.2 Historical Approaches to HLA Typing: Serology and DNA-RFLP Approaches

Early methods for HLA typing were based on the detection of expressed HLA molecules on the surface of separated T cells (HLA class I products) and B cells (HLA class II products) using panels of antisera, usually obtained from multiparous women in a complement-dependent cytotoxicity test (Mittal et al. 1968), commonly referred to as serological typing. While this approach is still commonly used in clinical laboratories worldwide – especially for HLA class I typing – it suffers from numerous deficiencies: viable lymphocytes are needed, the antisera required are generally non-renewable and the technique has limited powers of resolution, especially for HLA class II alleles. For example, of the 264 DRB alleles currently detectable at the DNA level (Bodmer et al. 1999), gene products from only 24 broad allele groups can be detected by serol-ogy. The same situation applies to HLA-DQB1, where only 6 out of 39 DNA sequence variants can be detected using allo-antisera. In addition, the com-plexity of the expressed epitopes recognised by allo-antisera leads to consid-erable antigen cross-reactivity, leading to a lack of operationally monospecific antisera for some HLA class II antigens. In addition, serological techniques are unable to detect HLA-DP specificities and are also of limited utility for HLA-C typing. While application of biochemical techniques such as one- and two-dimensional isoelectric focusing has helped to unravel certain serological ambiguities, these techniques are too cumbersome for widespread use.

Serological HLA typing methods are still used in some clinical laboratories and may have utility as a research tool, especially in the functional character-isation of antigens encoded by newly described HLA alleles (Hemmatpour et al. 1998) and in the identification of null alleles (e.g. Lienert et al. 1996). However, for the past decade, DNA-based methods have been the principal tools for HLA typing in both research and clinical laboratories. DNA methods offer practical advantages over serology in that viable cells are not required, reagents are renewable and there is increased potential for standardisation. In particular, the recent International Histocompatibility Workshops have been instrumental in encouraging development and implementation of new typing techniques and have provided a forum for their standardisation. The first comprehensive DNA-based HLA class II DR/DQ typing system was based on restriction fragment length polymorphism (RFLP) analysis (Bidwell et al. 1988). Briefly, this technique involved the use of restriction enzyme digestion of genomic DNA, followed by size separation of the resultant fragments by agarose gel electrophoresis, DNA fragment denaturation and Southern blotting onto a nylon membrane. These DNA fragments are detected by hybridisation with homologous labelled cDNA or DNA probes.

The identification of different alleles occurs according to their individual banding patterns, due to polymorphism in the DNA sequence targeted by the enzyme. Numerous early studies demonstrated the ability of RFLP typing to resolve serological ambiguities and mistypings (e.g. Howell et al. 1989). The clinical utility of this improved typing was also revealed early on. For example, the Collaborative Transplant Study showed an improvement in survival for renal allografts that had been DR-matched by RFLP in comparison with serology and a discrepancy of up to 25% in serological typing methods (Mytilineos et al. 1990; Opelz et al. 1993). Despite this success, RFLP is time-consuming (typing may take up to 10 days) and laborious. The method is limited since it is not related to sequence polymorphism in exon 2 (see Sect. 6.4.1), interpretation often relying on linkage disequilibrium between DR, DQ and neighbouring sequences, and it is difficult to apply to class I typing. In addition, most early methods relied on the use of radiolabelled probes. DNA-RFLP typing has now been almost entirely superseded by the more rapid polymerase chain reaction (PCR)-based methods, which are capable of much greater allelic resolution, rapidity and throughput, depending on the approach used. However, DNA-RFLP was of considerable significance as a forerunner of these techniques.

6.4 PCR-based Approaches to HLA Genotyping

6.4.1 Distribution of Polymorphisms Within HLA Genes

In developing strategies for PCR-based HLA typing, an understanding of the distribution of coding polymorphisms within the HLA class I and class II genes is essential. The basic intron/exon arrangement of the HLA class I and class II genes is shown in Fig. 6.2. The antigen-binding groove of class I molecules is encoded by exons 2 and 3 of the gene, while the class II antigen-binding groove is encoded by exon 2 of the respective A and B genes for that subregion. The vast majority of coding polymorphisms are contained within these exons. HLA-DR is a special case in that exon 2 of the DRA gene is not polymorphic.

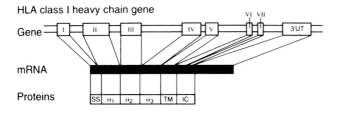

Fig. 6.2. Exon/intron structure of genes encoding HLA class I α (heavy chain) and class II α and β chains and the polypeptide domains of the HLA molecules encoded. The HLA class I peptide-binding cleft is composed of α1 and α2 domains and the class II peptide-binding cleft of α1 and β1 domains. *3′UT* 3′ Untranslated; *SS* signal sequence; *TM* transmembrane region; *IC* intracellular region (Krausa and Browning 1996)

In addition, polymorphism is clustered within a limited number of hypervariable regions within each exon (e.g. typically six hypervariable regions for exon 2 of an HLA class II gene such as DPB1). Polymorphic differences in the hypervariable regions of these exons are rarely allele-specific. Rather, hypervariable sequence motifs are often shared between a number of alleles and so a unique allele sequence is comprised of a unique combination of sequence motifs at these hypervariable regions. To more readily illustrate this, amino acid sequences encoded by the second exon of a number of DRB1 alleles are shown in Fig. 6.3. The situation with respect to HLA class I genes is more complex, not only because polymorphism is spread over two exons, but because sequence motifs at the hypervariable regions are not only shared between different alleles of the same locus, but also can be shared between alleles of more than one class I gene or pseudogene. As a result of this, the development of PCR-based methods for HLA class II typing has generally preceded that for class I. However, even for HLA class II, the presence of the same sequence motif in more than one allele complicates all DNA-based (and indeed, non-DNA-based) methods of HLA typing, necessitating the use of many reagents to identify each allele.

The number of reagents required may be very large when high-resolution (i.e. allelic) genotyping is required, or considerably smaller if initial typing for broad allelic groups (which may encode expressed molecules with consider-

```
             1       10        20        30        40        50        60        70        80        90        100
DRB1*0101    GDTRPRFLWQLKFECHFFNGTERVRLLERCIYNQEESVRFDSDVGEYRAVTELGRPDAEYWNSQKDLLEQRRAAVDTYCRHNYGVGESFTVQRRVEPKVTVY
DRB1*01021   ------------------------------------------------------------------------------------AV----------------
DRB1*01022   *****-------------------------------------------------------------------------------AV****************
DRB1*0103    ------------------------------------------------------------------I--DE-------------------------------
DRB1*0104    ----------------------------------------------------------------------------N--------V----------------
DRB1*15011   ----------P-R------------F-D-YF---------------F-------------------I---A--------------V---------Q------
DRB1*15012   *****-----P-R------------F-D-YF---------------F-------------------I---A--------------V-------*********
DRB1*15021   ----------P-R------------F-D-YF---------------F-------------------I---A------------------------Q------
DRB1*15022   *************------------F-D-YF---------------F-------------------I---A---------------****************
DRB1*15023   *****-----P-R------------F-D-YF---------------F-------------------I---A---------------****************
DRB1*1503    ----------P-R------------F-D-HF---------------F-------------------I---A--------------V---------Q--****
DRB1*1504    ******----P-R------------F-D-YF---------------F-------------------F---A--------------V-----***********
DRB1*1505    ******----P-R------------F-D-YF---------------F-----------------------A--------------V-------*********
DRB1*1506    ********--P-R------------F-D-YF---------------F--A----------------I---A--------------V-***************
DRB1*16011   ----------P-R------------F-D-YF-----------------------------------F--D-------------------------Q------
DRB1*16012   *****-----P-R------------F-D-YF-----------------------------------F--D--------------------************
DRB1*16021   ----------P-R------------F-D-YF--------------------------------------D-------------------------Q------
DRB1*16022   ******----P-R------------F-D-YF--------------------------------------D----------------****************
DRB1*1603    ----------P-R------------F-D-YF-----------------------------------F--DA------------------------Q------
DRB1*1604    ********--P-R------------F-D-YF-----------------------------------F--D---L------------****************
DRB1*1605    *******---P-R------------F-D-YF-----------------------------------I--D-----------------***************
DRB1*1607    *****-----P-R------------FPD-YF-----------------------------------I--D---------------------***********
DRB1*1608    *****-----P-R------------F-D-YF-----N-----------------------------F--D------------------------********
DRB1*03011   --------EYSTS------------Y-D-YFH----N---------F-----------------------K-GR--N--------V---------H------
DRB1*03012   *****---EYSTS------------Y-D-YFH----N---------F-----------------------K-GR--N--------V------**********
DRB1*03021   --------EYSTS------------F---YFH----N---------------------------------K-GR--N------------------H------
DRB1*03022   --------EYSTS------------F---YFH----N---------------------------------K-GR--N------------------H----**
DRB1*0303    *********YSTS------------F---YFH----N---------------------------------K-GR--N--------V--------********
DRB1*0304    *******-EYSTS------------Y-D-YFH--------------F-----------------------K-GR--N--------V-***************
DRB1*0305    *****---EYSTS------------Y-D-YFH----N---------F-----------------------K-GR--N-----------------********
DRB1*0306    ******--EYSTS------------Y-D-YFH----N---------------------------------K-GR--N--------V------**********
DRB1*0307    --------EYSTS------------F-D-YFH----N---------F-----------------------K-GR--N--------V---------H------
DRB1*0308    --------EYSTS------------Y-D-YFH----N---------F----------E------------K-GR--N--------V---------H------
DRB1*0309    *******-EYSTS------------Y-D-YFH-R--N---------F-----------------------K-GR--N----------------*********
DRB1*0310    --------EYSTS------------Y-D-YFH----N---------F---------A--H----------K-GR--N--------V---------H------
DRB1*0311    *****---EYSTS------------Y-D-YFH----N---------F-----------------------K-GQ--N--------V--------********
DRB1*04011   --------E-V-H------------F-D-YF-H---Y---------------------------------K------------------------Y-E----
DRB1*04012   *****---E-V-H------------F-D-YF-H---Y---------------------------------K-----------------------********
DRB1*0402    --------E-V-H------------F-D-YF-H---Y-----------------------------I--DE--------------V--------********
DRB1*0403    --------E-V-H------------F-D-YF-H---Y------------------------------------E-----------V---------Y-E----
DRB1*0404    --------E-V-H------------F-D-YF-H---Y------------------------------------------------V---------Y-E----
DRB1*04051   --------E-V-H------------F-D-YF-H---Y-------------------S-------------------------------------********
DRB1*04052   ******--E-V-H------------F-D-YF-H---Y-------------------S-------------------------------------********
```

Fig. 6.3. Examples of HLA-DRB1 amino acid sequences encoded by exon 2 sequences

able functional similarity) is sufficient for a primary screen. Thus most PCR-based HLA typing methods contain an inherent tension between resolution and throughput. Excluding direct sequencing and DNA conformation-based methods (Sects. 6.4.6 and 6.4.7), hypervariable motif-directed reagents include restriction enzymes, sequence-specific oligonucleotide probes (SSOPs) or oligonucleotide primer pairs for sequence-specific (SSP) PCR. These methods and their advantages and disadvantages for genetic epidemiological studies will be considered in turn.

6.4.2 DNA Extraction Methods

Availability of template DNA of suitable quality and quantity is a prerequisite for successful HLA typing by PCR. Currently, a wide range of DNA extraction methods is available, including standard phenol-chloroform mini- and bulk-preps and rapid salt precipitation methods (Miller et al. 1988), from a variety of starting materials, including peripheral blood, mouthwashes (Aron et al. 1994) and formalin-fixed, paraffin-embedded archival histopathological biopsy tissues (Howell et al. 1995). It is beyond the scope of this review to detail these various methods. However, the question of starting material and method-ology for DNA preparation requires considerable attention when planning a study, since some methods require more DNA per successful typing or are sen-sitive to DNA quality. In general DNA extracted from archival biopsy tissues is variably and often extensively degraded, limiting amplicon length for success-ful PCR. In our experience, amplicon length should not exceed 300 bp if reli-able PCR is to be achieved. A DNA extraction method for such fixed tissues is given in Section 6.5.2. Ethical issues also arise when the use of such tissues for retrospective studies is proposed, which require careful consideration.

6.4.3 PCR-RFLP

Although restriction enzymes were originally used for HLA class II DR/DQ typing by direct RFLP analysis, the utility of this technique is limited (Sect. 6.3.2). However, when combined with initial PCR amplification of the gene or exon in question and based on knowledge of the allele sequences in question, restriction enzyme cutting sites can be identified within PCR amplicons. Therefore, HLA typing can be performed by digesting PCR products with a number of restriction enzymes, and this approach has successfully been applied to both HLA class I and class II typing (e.g. Uryu et al. 1990; Tatari et al. 1995). Despite being a fairly rapid method (4–5 h), which does not require the use of radioisotopes, PCR-RFLP methodology can be complex and expen-sive, requiring the use of multiple restriction enzymes. In addition, interpre-tation of results can be problematic, with partial restriction digestion being a particular problem (Bidwell 1992) and the continued rapid expansion in the numbers of known HLA alleles has limited the widespread application of this methodology, which is therefore unsuitable for genotyping large numbers of samples.

6.4.4 PCR-SSOP Typing

The principle of this widely used approach is that individual alleles or allele groups can be discriminated by hybridisation of the immobilised PCR product containing the hypervariable regions to be typed with appropriately labelled SSOPs. For HLA class II typing by PCR-SSOP, PCR amplication of the entire exon 2 sequence must be performed, while for PCR-SSOP typing of HLA class I genes the PCR primers flank exons 2 and 3. Primers must be complementary to sequences that are common to all alleles to be differentiated. The resultant PCR product is denatured and then immobilised on a series of nylon membranes, and probed with a series of labelled oligonucleotides, each of which is complementary to a different sequence motif at each of the hypervariable regions. This is referred to as the "dot-blot" method. A single dot blot may be serially interrogated with a series of SSOPs or a series of dot blots can be produced and each dot blot hybridised with a separate SSOP. Under stringent washing conditions, the probes will only remain bound to exactly complementary target sequences and bound probes can be detected by an appropriate visualisation method, e.g. autoradiography for radiolabelled oligonucleotide probes, which have largely been superseded by digoxigenin- or biotin-labelled probes, which can be detected by colourimetric or more usually enhanced chemiluminescence methods. A panel of probes can identify which polymorphisms, and consequently which alleles, are present by the pattern of positive reactions. For example, a limited panel of 18 probes can be used for successful low-resolution typing of the DRB1 locus, with a resolution still much superior to that of good serology. A panel of 76 probes in conjunction with 5 separate allele group-specific PCR reactions has been recommended for allelic HLA-DRB typing (Fernandez-Vina and Bignon 1997). General principles of SSOP design, validation, hybridisation and stringency washing are dealt with in Chapter 5 (this Vol.).

Theoretically, PCR-SSOP can detect any allele of any HLA locus, as long as the sequence is known. This approach was initially applied to HLA class DR, DQ and DP typing (Kimura and Sasazuki 1992), but has since been extended to the HLA class I A, B and C loci (Kennedy et al. 1997) and the "non-classical" TAP, LMP and HLA-DM loci (Powis and Teisserenc 1997). It is now used for HLA typing in a variety of fields, such as forensics, disease association studies and matching for transplantation (reviewed by Krausa and Browning 1996), and can be applied using varying numbers of probes to achieve the allelic resolution required. Up to 96 samples can be dotted manually on a single membrane and, using automated equipment, this number can be doubled or quadrupled, allowing a relatively rapid throughput of samples in comparison to RFLP. Therefore, it represents a valuable technique for non-urgent large-scale typing, for instance bone marrow donor panels or epidemiological studies. However, it is less suitable for smaller numbers of samples, as a complete panel of probes, with a separate dot-blot membrane for each, is required, making the procedure very laborious. In addition, data analysis can often be complicated, particularly with some heterozygotes, since this approach cannot distinguish between sequence motifs present in *cis* or *trans*. Some of these combinations can only be distinguished using allele group-specific PCR reactions and subsequent SSOP typing, or via the application of another approach. In addition, a crucial step in the procedure is accurate control of temperature during the stringent washing of membranes, if ambiguous hybridisation signals are to be avoided. It should also be noted that a series of control PCR products must always be included on all membranes, so as to give at least one positive and one negative hybridisation signal for each probe used. A considerable drawback of this approach is the time required for manual recording and interpretation of dot-blot results. Automated data gathering of positive and negative probe hybridisations and interpretation of these patterns would constitute a major advance with respect to high-resolution genotyping in epidemiological studies, although this is less critical in most clinical laboratories.

A variation on the PCR-SSOP direct dot-blot method is the reverse dot blot, where it is the oligonucleotide probes which are immobilised on a single membrane, and the PCR product which is labelled and hybridised in solution. This method has been extended to include the use of immobilised probes in microtitre plates (Giorda et al. 1993) and has the advantage of rapid (3–4h) analysis of small numbers of samples. Application of microchip technology would also be beneficial, although expensive. The main disadvantage of this technique is still the complex interpretation required, unless automated recording and interpretation of positive hybridisation signals can be implemented.

6.4.5 PCR-SSP or ARMS-PCR Methods

This approach is based on the original discovery that specificity in a PCR reaction is dependent on precise matching of the terminal 3′ end of a PCR primer and its target DNA sequence. Using this approach, point mutations may be distinguished from wild-type sequences using a single generic primer combined with one of two antisense primers in separate PCR reactions. One of the antisense primers is precisely matched with the wild-type sequence at its 3′-end, while the second primer is precisely matched with the mutant sequence at it's 3′-end. Therefore, one PCR reaction will only amplify the wild-type sequence and the other will only amplify the mutant sequence. This method, originally termed the "amplification refractory mutation system-PCR" (ARMS-PCR)

(Newton et al. 1989), is dependent on the fact that Taq DNA polymerase lacks 3'- to 5'-exonucleolytic proof-reading activity, so that Watson-Crick mismatches at the 3'-end of the primer-template duplex cannot be corrected, which would result in mispriming. Successful application of the ARMS-PCR approach also requires stringent conditions for primer annealing in the PCR reaction. A vital component of this methodology is inclusion of a second primer pair which amplifies a sequence from a second gene, to act as an in-tube positive control for successful or failed PCR amplification in that tube.

The first comprehensive approach to HLA-DRB1 genotyping using this method was developed in 1992 by Olerup and Zetterquist. These authors renamed this approach "PCR with sequence-specific primers" (PCR-SSP). A comprehensive method for combined HLA class I A, B, C and class II DRB1, DRB3, DRB4, DRB5 and DQB1 genotyping has since been developed (Bunce et al. 1995). In all such methods, primer pairs are combined into allele- or, more usually, allele-group-specific PCR reactions, each containing an internal control PCR primer pair. While some of the reactions are "single ARMS" in that specificity is due to a single primer, in most of the PCR reactions specificity is conferred by the specific combination of both sense and antisense primers, i.e. "double ARMS". Panels of allele-specific primer mixes are used for each locus to be typed. Primer mixes can be pre-aliquotted and dried in 96- or 192-well microtitre plates for ease of use. HLA genotype is interpreted from the pattern of positive and negative PCR reactions for the locus in question. In usual laboratory practice, successful PCR amplification is visualised by ethidium bromide agarose gel electrophoresis, UV transillumination and photography.

The number of primer mixes required for PCR-SSP typing depends on the degree of resolution required. In our laboratory, low-resolution HLA (i.e. allele group) typing is performed using 31 reactions for DRB/DQB1, and 96 reactions for HLA-A, B and C. Subsequent high-resolution (i.e. allelic) typing can then be used to discriminate individual alleles within a broad allele group using additional PCR reactions. High-resolution typing is performed by the same basic SSP method, and interpretation is generally straightforward. However, in some cases – especially when both alleles are of the DRB1*04 subgroup – ambiguous heterozygous combinations occur, which cannot be resolved by a single round of PCR-SSP and a nested PCR-SSP approach is required. PCR-SSP analysis using the standard PCR amplification conditions developed for HLA typing has also been applied for genotyping other loci under investigation in immunogenetics laboratories, including the human platelet antigen genes (Metcalfe and Waters 1993), ABO and Lewis blood groups (Proctor et al. 1997), HFE (Guttridge et al. 1998) and various cytokine promoter and regulatory polymorphisms (Perrey et al. 1999). In addition, we have developed a nested PCR-SSP method for HLA class II DRB1 typing very limited amounts of DNA, e.g. from small biopsy tissue fragments (Bateman et al. 1997).

The main advantage of PCR-SSP over PCR-SSOP is that it is a relatively straightforward procedure which can be used to type a single sample in less than 3 h. Thus it is a valuable technique for HLA class I and class II typing in the clinical laboratory, especially for the typing of cadaveric organ donors. In addition, the use of an ARMS or double ARMS approach avoids many of the problems of resolving problematic heterozygotes, because successful amplification can only occur if the hypervariable sequence motifs targetted by the sense and antisense primers are present within the same allele. However, the use of large numbers of separate PCR reaction mixes for each sample to be typed renders this approach unsuitable for typing large numbers of samples. In addition, with the discovery of ever-increasing numbers of alleles, high-resolution HLA genotyping may require increased reliance upon nested PCR-SSP-based approaches. PCR-SSP generally requires more DNA of a higher quality than does PCR-SSOP, which may make the latter technique more readily applicable to whole genome amplified DNA.

Further modifications have been made to the standard SSP method. While almost all systems rely on multiple single allele or allele-group-specific PCR reactions, multiplex PCR for low-resolution HLA-A typing has been developed (Browning et al. 1994). However, to date, this approach has not been widely used, limited in part by sharing of polymorphic sequence motifs between multiple alleles of the same locus. More promisingly, in order to reduce post-PCR processing, fluorotyping systems have been developed. These utilise a fluorescent dye, e.g. ethidium homodimer (Ferencik and Grosse-Wilde 1993) or yellow oxazol (Bein et al. 1994), which intercalates with the PCR product, allowing direct detection without the need for agarose gel electrophoresis. The main drawback with this technique is that a positive PCR control cannot be used. However, Luedeck and Blasczyk (1997) have described a fluorotyping system for HLA-C that allows the detection of both the HLA allele and a control PCR product, by the use of two different dyes and a "TaqMan" approach. A similar method has also been described for DQB1 (Faas et al. 1996).

6.4.6 Conformation-Based Methods

Unlike the above methods, which all depend on the identification of multiple polymorphic sequence motifs within one or more exons of the HLA gene in question, a number of methods for HLA typing or determining HLA identity/non-identity between individual DNA samples have been developed, based on the conformation and mobility of single-stranded or double-stranded PCR products when subjected to polyacrylamide gel electrophoresis (PAGE). These methods will be considered briefly, since evolution of the basic methodology has resulted in development of reference strand conformational analysis (RSCA) genotyping (Sect. 6.4.6.3), which has considerable potential for rapid, allelic genotyping over a range of throughputs.

6.4.6.1 PCR with Single-Strand Conformation Polymorphism (PCR-SSCP)

PCR-SSCP relies on the fact that in single-stranded form PCR products containing different sequence polymorphisms will adopt different conformations, and will therefore differ in mobility in non-denaturing PAGE. These unique mobilities, which arise as a result of intrastrand complementary base pairing, can be used to identify particular allelic products. This approach has been applied successfully for both HLA class I and class II typing (e.g. Shintaku et al. 1993; Blasczyk et al. 1995) but is a relatively complex technique and results can be subject to interpretation difficulties. Accordingly, PCR-SSCP has not been applied widely for HLA genotyping. Even when technically simplified or partly automated (Kimura et al. 1993), these problems remain and PCR-SSCP is therefore not recommended for genetic epidemiological studies. However, this method is particularly useful for screening for new polymorphisms and has played a role in the detection of novel alleles.

6.4.6.2 Heteroduplex Analysis

Heteroduplex analysis (HDA) represents the forerunner of RSCA. HDA is a conformation-based technique that relies on the formation of mismatched heteroduplexes between closely similar but non-identical complementary DNA strands during PCR. These strands may arise from co-amplification of sequences from different alleles of the same gene in heterozygotes, from closely related genes (e.g. DRB1, DRB3, DRB4 and DRB5) in homozygotes and heterozygotes or from co-amplification of the same gene sequences in DNA admixtures from individuals non-identical for the gene sequence in question. In the latter situation, HDA or "DNA cross-matching" has been used to HLA class II match patients with unrelated bone marrow donors (Clay et al. 1991). In this simple form, the HDA technique is limited by its inability to define alleles. However, addition of an artificial construct or "universal heteroduplex generator" to PCR reactions can lead to the generation of allele-specific heteroduplexes and in some instances this can be applied for HLA genotyping, e.g. for HLA-DRB1*04 subtyping (Savage et al. 1996). However, despite the relative cheapness of this approach, it does not allow for comprehensive HLA genotyping and is subject to a number of interpretation difficulties. It is therefore unsuitable for application to large-scale genetic studies.

6.4.6.3 Reference Strand Conformational Analysis (RSCA)

More recently, a conformational technique which utilises a "fluoroscein-labelled reference" (FLR) has been described in relation to HLA class I A, B and C typing (Arguello et al. 1998). This technique has also been extended to DPB1 genotyping (Ramon et al. 1998), while DRB/DQB1 systems are under development (A.-M. Little, pers. comm.). Duplexes are formed between the PCR product of the locus of interest and a locus-specific FLR strand. A unique heteroduplex is formed between the labelled FLR strand and the complementary strand from the sample DNA, as all the detectable duplexes have the FLR in common. Thus the mobility of the duplex is determined by the allele present. PCR products are separated by PAGE and mobilities determined with an automated DNA sequencer. Labelled duplexes are detected using a laser detection system, and computer software is used to create a "mobility scale" for each allele. In this manner, Arguello et al. (1998) blindly typed over 250 samples and in so doing identified a new variant of HLA-A*0301. RSCA is potentially a rapid method for high-resolution typing of large (or small) numbers of samples, but it is difficult to establish for HLA genotyping without an extended development period or the purchase of newly available RSCA typing kits, as yet only available from a single supplier (Pel-Freez) and optimised with software appropriate for one manufacturer's DNA sequencer. Despite this, RSCA will undoubtedly play a considerable role in future large-scale DNA typing strategies.

6.4.7 Sequence-Based Typing

Despite their tremendous utility, PCR-SSOP- and PCR-SSP-based methods can only test for known HLA polymorphisms, although aberrant typing results may provide preliminary evidence for potential new alleles (e.g. Hemmatpour et al. 1998). Ultimately, direct DNA sequencing is needed to identify and characterise such alleles and the full extent and complexity of HLA polymorphisms will remain unknown without the widespread application of DNA sequencing technology to the HLA complex. However, sequencing is still a comparatively cumbersome procedure for analysing large numbers of samples and, in the HLA community, sequenced-based typing (SBT) is largely confined to identification of potential novel alleles. However, continual improvements in sequencing technology, combined with the availability of commercial software for interpretation of sequences and assignments of genotypes by PCR-based sequencing of DNA direct from heterozygous individuals, may permit the more widespread application of sequencing for larger-scale HLA typing (Bettinotti et al. 1997). However, prior separation of alleles by group-specific PCR may still be necessary prior to sequencing for some heterozygotes, to reduce the number of ambiguous genotypes which may be generated by automated sequencing, adding to the complexity of this approach. In the case of HLA, where multiple polymorphic sites occur in the same gene, simple sequencing of locus-specific PCR products suffers from the same problem of resolving certain heterozygotes as does PCR-SSOP.

6.4.8 HLA Genotyping Techniques: Out of Many, One?

As the above summary reveals, many different PCR-based HLA typing techniques are in current use, often in the same laboratory, which may appear both wasteful and unnecessary. Could much effort be saved if a single method, capable of allelic genotyping, were adopted as standard? However, there are a number of reasons as to why – as yet – this has not been so. The technique selected for a particular application depends on a number of parameters including the degree of allelic resolution that is required, the number of samples to be typed, time constraints (if any) and, of course, economic considerations. For the majority of applications, a medium resolution method is usually the best starting point. The researcher can then focus upon candidate groups of alleles that can be investigated further using higher-resolution techniques. In addition, not all techniques discriminate between all combinations of alleles. A secondary technique is needed to provide additional discrimination. Application of more than one technique in the same laboratory also provides added quality control for the primary typing method used. Furthermore, in the case of PCR-SSOP typing, inclusion of all probes necessary for allelic typing of a single locus would be particularly cumbersome and expensive. While microchip technology would reduce the first limitation, set-up costs would be further increased. In general, PCR-SSOP and PCR-SSP methods cover a wide range of applications and can be adapted to provide a wide range of resolution. PCR-SSOP is the preferred method for high-throughput, high-resolution HLA genotyping and, while a "traditional" technique, it has considerable scope for automation. PCR-SSP is rapid, but much less readily adapted to the needs of high-throughput analyses, particularly when high-resolution typing is required.

Nevertheless, automation of PCR-SSP reaction mix preparation in microtitre plates and direct detection of PCR products avoiding the need for agarose gel electrophoresis is making this approach more attractive to the researcher. Alternatively, if PCR-SSP typing were combined with microplate-array diagonal gel electrophoresis (MADGE), throughput could be increased by 10–100-fold, enabling a single user to run up to several thousand gel lanes per day (Day et al. 1998, 1999). As yet, the potential for multiplexing PCR-SSP typing remains largely unfulfilled. While the principal use of PCR-SSP HLA typing is in the clinical laboratory, PCR-SSP can play a useful role in PCR-SSOP-based approaches, when used selectively to resolve ambiguous heterozygotes. Finally, typing methods based on conformational analysis have one important advantage in that amplicon sequences not directly targeted by PCR primers or SSOPs can be examined. In addition, as the numbers of known alleles increase, unlike PCR-SSP or PCR-SSOP, there is usually no need to design additional primers and probes. Conformational methods such as PCR-SSCP and RSCA can also be used to screen for new alleles, which can be confirmed by DNA sequencing.

6.5 A PCR-SSOP-Based Strategy for HLA Class II Typing Using DNA Derived from Archival Tissue Banks

6.5.1 Introduction

In our laboratory we have a particular interest in retrospective studies of HLA-disease associations, using archival histopathological biopsy tissues as a source of DNA for analysis. This approach has been used to investigate the immuno-genetics of the evolution of enteropathy-associated T cell lymphoma within pre-existing coeliac disease (Howell et al. 1995), HLA-mediated genetic susceptibility to cutaneous malignant melanoma (Bateman et al. 1998), and for the identification of contamination of diagnostic histopathological biopsies with tissue from extraneous sources (Bateman et al. 1994). We have also developed PCR-SSP-based HLA class II genotyping approaches for application to similar material (Bateman et al. 1996, 1997) but, for the reasons outlined in Section 6.4.8, PCR-SSP is less suitable for high-throughput HLA genotyping studies, although if combined with MADGE approaches, PCR-SSP would be less constrained. While utilisation of archival biopsy tissue for genetic studies is an attractive option, since patient clinical and pathological follow-up data of many years duration may be available, DNA obtainable from such biopsies is generally and variably degraded and less DNA is recoverable from some tissues than others; e.g., in our experience, we can only recover limited amounts of DNA from cardiac biopsies. Accordingly, a strategy is presented for medium-resolution HLA class II genotyping, using formalin-fixed, paraffin wax embedded tissue as starting material. This method can be expanded into a high-resolution typing system using additional PCR primers and SSOPs, as outlined by Fernandez-Vina and Bignon (1997), and also applied to DNA extracted from peripheral blood or fresh tissues.

6.5.2 DNA Extraction

Five 10-μm sections are cut from each tissue block and placed in a 1.5-ml Eppendorf tube and 1 ml of xylene added. Tubes are vortexed for 1 min and then centrifuged at 13,000 g for 5 min. The supernatant is discarded and a further 1 ml of xylene is added to the pellet which is then re-vortexed and centrifuged. Again, the supernatant is discarded and 1 ml of a 1:1 xylene/absolute ethanol mixture is added to the pellet. This is then vortexed and centrifuged as above and the supernatant discarded. The xylene/ethanol extraction step is repeated and, following this, 1 ml of absolute ethanol is added to the pelleted material, vortexed and centifuged as above. This ethanol extraction step should be repeated and the supernatant discarded and all traces of ethanol removed, followed by air incubation at 37 °C for 20 min. Lysis buffer (100 μl; 100 mM Tris pH 8.0, 4 mM EDTA, 0.45% NP40, 0.45% Tween 20) containing 6.7 μl of 6 mg/ml proteinase K is added to the pellet. The volume of lysis buffer can be increased up to 200 µl if the pellet is large. This extraction stage should be incubated overnight at 55 °C to achieve maximum DNA yield. Following this incubation, proteinase K is inactivated by heating at 100 °C for 10 min. The sample is then spun at 13,000 g for 5 min and the DNA-containing supernatant drawn off. This DNA can then be used directly as a template in PCR reactions.

6.5.3 PCR Amplification

HLA class II DRB, DQA1, DQB1 and DPB1 typing is achieved by PCR amplification of the second (variable) exon of the class II gene in question using primers derived from the 11th International Histocompatibility Workshop (Kimura and Sasazuki 1992). Five µl of DNA are used per 100 µl of PCR reaction mix, which comprises 1.5 mM $MgCl_2$, 12% sucrose, dNTPs to a final concentration of 200 µM for each dNTP, each PCR primer at a concentration of 0.2 µM, 1 unit Taq DNA polymerase and enzyme manufacturer's buffer. Following an initial denaturation step of 94 °C for 5 min, the PCR programme is as follows: denaturation at 94 °C for 30 s, annealing at 61 °C for 1 min and extension at 72 °C for 1 min 30 s for 30 cycles. A final elongation step of 72 °C for 5 min completes the PCR programme. Following PCR, 10 µl of PCR product is run on a 2% agarose gel, stained with ethidium bromide and examined under UV transillumination to assess PCR product yield.

It should be noted that a large-volume PCR reaction is required to generate a sufficient volume of PCR product for the preparation of multiple dot blots per locus to be typed.

6.5.4 Dot-Blot Preparation

At this stage in the procedure, a small amount of PCR product is spotted onto a positively charged nylon membrane, denatured and the resulting single-strand DNA is immobilised by UV cross-linking, prior to hybridisation with a labelled oligonucleotide probe. Dot blots can be prepared by manual addition of PCR products to a nylon membrane, in a 96-sample array, or a 96-well replicator may be used. Ninety-six or 192-channel multiple dispense instruments (Robbins Scientific) can also be used for the preparation of replicate dot blots, achieving a considerable time saving over manual methods. For ease of subsequent interpretation, whatever approach is used for spotting PCR products onto membranes, it is helpful if approximately equal amounts of PCR products are used per individual spot on a given membrane.

For manual preparation of dot blots, PCR reaction products are heated at 95 °C for 5 min to denature the PCR product. Products are then chilled on ice for 10 min prior to spotting onto replicate membranes. Usually 2–3 µl of PCR product per spot will suffice, but if PCR has only generated a weak product (determined by prior agarose gel electrophoresis of PCR reaction products), spots can be allowed to dry and a further aliquot spotted on top of the first (such adjustments are much more difficult if an automated dispense system is used for dot-blot preparation). Spots should be laid out in a "96-well" array, or a "192-well" array if an automated 192-channel multidispense instrument is used. Complete denaturation of DNA is ensured by wetting the membrane in denaturation solution (1.5 M NaCl, 0.5 M NaOH) for 5 min, by placing membranes on a wad of filter paper soaked in the solution. Membranes are then transferred to neutralising solution (1.5 M NaCl, 1 mM EDTA) for 1 min and dried by blotting on filter paper and air dried. DNA is immobilised on membranes by wrapping in SaranWrap and UV irradiating for a pre-optimised period (usually around 2 min, depending on the output of the transilluminator). Replicate membranes should be prepared, ideally one per SSOP to be used, but if large panels of SSOPs are to be used for high-resolution typing, this may not always be practical and some membranes will need to be dehybridised and re-probed with further SSOPs. Each membrane must contain at least one positive and one negative hybridisation control for each SSOP used.

6.5.5 Dot-Blot Prehybridisation, Hybridisation and Stringency Washes

Membranes are individually placed in 50-ml plastic centrifuge tubes and incubated for 30 min at room temperature in 10 ml blocking solution [4 × SSPE, 1% blocking reagent (Boehringer, Mannheim), 0.1% lauroyl sarcosine]. Following this, membranes are prehybridised in 5 ml tetramethylammonium chloride (TEMAC) hybridisation buffer (3 M TEMAC, 50 mM Tris, pH 8.0, 2 mM EDTA, 0.1% SDS) for 50 min at 54 °C. Approximately 2 pmol/ml of digoxigenin-labelled SSOP is added to the prehybridisation solution and hybridised for 75 min at 54 °C. For medium-resolution HLA class II typing, 18 DRB1 (SSOPs 1001–1005, 1007, 1008, 2803, 3703, 3709, 5703, 5704, 7001, 7004, 7007, 7009, 7030 and 8602), 10 DQA1 (2502, 3401–3403, 4102, 5501–5504 and 6903), 18 DQB1 (2301, 2302, 2601–2604, 2606, 3702, 3703, 4501, 4901 and 5701–5708), and 19 DPB1 (0901–0904, 3503, 5501–5504, 6502, 6901–6903, 6905, 6906, 7601, 7602, 8501 and 8503) SSOPs are used, selected from the 12th International Histocompatibility Workshop panel (SSOP numbers in brackets refer to those described by Fernandez-Vina and Bignon 1997). All SSOPs are the same length (18mer), except for DPB1 6903, which is a 20mer.

Prehybridisation and hybridisation steps are performed with continuous rotation of tubes in a dedicated hybridisation oven (e.g. Hybaid, Robbins Scientific). Following hybridisation, membranes are removed from the centrifuge tubes and placed in a plastic box for subsequent stringency washes to remove incompletely matched SSOP-template duplexes. Hybridisation solutions containing labelled SSOPs are not discarded, but are stored at –20 °C for subsequent re-use. Solutions can be used up to five times, with addition of very small amounts of extra probe, if required.

Stringency washes are performed as follows: two 5-min washes in $2 \times$ SSPE (saline sodium phosphate/EDTA), 0.1% SDS, followed by two 10-min washes in 3 M TEMAC, 50 mM Tris pH 8.0, 2 mM EDTA, 0.1% SDS at 58 °C for all probes except DPB1 6903 (58 °C is a stringent temperature for an 18mer). Final washing should be performed at 60 °C for DPB1 6903. The use of probes of identical length, combined with TEMAC-based washes, allows for uniform stringency wash conditions, unless indicated otherwise.

6.5.6 Detection of Bound SSOPs

The purpose of this stage in the procedure is to visualise SSOPs which are bound to single-stranded immobilised PCR products by enhanced chemiluminescence and to obtain a permanent record of results by autoradiography (or, more properly, "luminography"). Since digoxigenin-labelled SSOPs are used in this genotyping procedure, some of the detection reagents used are proprietary products available from Boehringer, Mannheim. All stages until incubation to initiate the chemiluminescent reaction are performed at room temperature.

Firstly, membranes are rinsed in "buffer 1" (1 M Tris, pH 7.5, 1.5 M NaCl) for 5 min. Following this, membranes are transferred to glass hybridisation bottles (Hybaid) and 20 ml of "buffer 2" [0.1 M Tris, pH 7.5, 0.15 M NaCl, 1% blocking reagent (Boehringer, Mannheim)] added and incubated for 30 min. This step is to prevent subsequent non-specific binding of anti-digoxigenin fab antibody fragments coupled with alkaline phosphatase (anti-DIG AP; Boehringer, Mannheim). Anti-DIG AP fab fragments (2 µl) are then added to 10 ml fresh "buffer 2" and membranes are incubated in this for 30–40 min. Membranes are removed from the hybridisation bottles, placed in a plastic box and unbound antibody conjugate washed off by three washes with 500 ml "buffer 1" for 10 min per wash. Membranes are then equilibrated in 200 ml "buffer 3" (0.1 M Tris, pH 9.5, 0.1 M NaCl, 50 mM $MgCl_2$) for 5 min.

The following steps visualise bound probe. Firstly, 40 ml of diluted reagent CSPD (Boehringer Mannheim) are prepared by dilution 1:100 in "buffer 3". CSPD is dephosphorylated by anti-DIG AP to produce an unstable intermediate that emits visible light as it breaks down. In a darkroom, membranes are incubated for 5 min in diluted CSPD. Diluted CSPD is then poured off and, if stored in the dark at 4 °C, can be re-used the next day by adding a further 20 µl CSPD stock. Membranes are blotted dry, wrapped in SaranWrap, sealed in a plastic bag and incubated at 37 °C for 15 min to initiate the chemiluminescent reaction. Following this, membranes are taped into an X-ray cassette and exposed to X-ray film. Two films are placed in an autoradiography cassette and the first film developed after 20 min to assess the optimum exposure period for the second film (usually an additional 10–20 min). An example

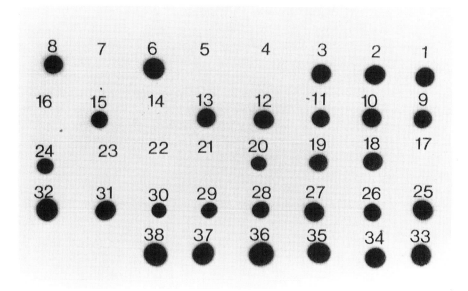

Fig. 6.4. PCR-SSOP dot-blot results obtained with the DPB1 SSOP number 8503, to illustrate clear positive and negative probe hybridisations detected by chemiluminescence

of a PCR-SSOP dot blot generated using DPB1 SSOP number 6903 is given in Fig. 6.4.

6.5.7 Dot-Blot Interpretation and Assignment of Genotypes

For each dot blot and SSOP, positive and negative control hybridisations are examined first to ensure that each SSOP is giving clear positive and negative hybridisation signals when expected. For each membrane, all other sample hybridisations are scored using an in-house numerical system equating to "definite positive", "probable positive", "equivocal", "probable negative" and "definite negative". From this inspection, when results with all SSOPs are compared, definite positive and negative hybridisations can be assigned and these results used to assign genotypes from the known sequence motif specificity of each SSOP. These assignments can be made manually, or using one of a number of simple computer programmes. Some workers have developed image gathering software which can be coupled to similar programmes, so allowing automated interpretation of dot-blot hybridisation patterns (G.M. Taylor, pers. comm.). Since data interpretation can be a rate-limiting step in PCR-SSOP genotyping, automated data interpretation and genotype assignment lead to a considerable increase in typing throughput.

6.5.8 Dehybridisation and Reprobing of Dot Blots

Following dot-blot SSOP hybridisation and detection, bound SSOPs can be removed and dot blots probed with further SSOPs. This approach can be useful if insufficient PCR products are available for the preparation of one membrane per SSOP. For example, if 20 SSOPs are required for a given genotype, ten membranes could be prepared and each probed with two separate probes. In some circumstances, such an approach may be preferable logistically. In addition, if results from a preliminary analysis of one locus suggest that genotyping at an increased resolution is indicated, membranes may be probed with an additional panel of selected SSOPs. Dehybridisation is simply achieved by subjecting membranes to three 10-min washes in 0.4 M NaOH at 45°C, followed by three 10-min washes in 0.2 M Tris, pH 7.5, 0.1 × SSC, 0.1% SDS at 45°C.

6.6 Concluding Remarks

The past 10 years have witnessed a continuing revolution in the application of DNA-based techniques for HLA genotyping, driven both by clinical and research needs. Use of these techniques has revealed an ever-increasing number of HLA alleles, with more than 225 new alleles described over the past 3 years (Bodmer et al. 1999). A wide variety of techniques are currently in use for detecting HLA polymorphism, with no single technique applicable to all research or clinical applications. Allelic resolution and the required throughput for a given study will critically determine which approach is selected. Traditional methodologies are still adequate for small-scale research projects, but automation of traditional technologies, including data gathering and allele calling, or new approaches such as RSCA are required for large-scale investigations involving thousands of subjects. The HLA system, with a number of closely related genes, multiple polymorphic sequence motifs within single genes and sharing of these motifs between alleles of the same locus and indeed between loci, has provided a demanding testing ground for molecular genotyping strategies capable of unravelling this complexity, which may therefore be applied with relative ease to less complex loci.

References

Amiel JC (1967) Study of the human leukocyte phenotypes in Hodgkin's disease. In: Dausset J, Colombani J (eds) Histocompatibility testing. Munksgaard, Copenhagen, pp 79–81
Arguello JR, Little A-M, Pay AL, Gallardo D, Rojas I, Marsh SGE, Goldman JM, Madrigal JA (1998) Mutation detection and typing of polymorphic loci through double-strand conformation analysis. Nat Genet 18:192–194

Aron Y, Swierczewski E, Lockhart A (1994) A simple and rapid micromethod for genomic DNA extraction from jugal epithelial cells. Allergy 49:788–790

Bateman AC, Howell WM (1999) Human leukocyte antigens and cancer. J Pathol 188:231–236

Bateman AC, Leung ST, Howell WM, Roche WR, Jones DB, Theaker JM (1994) Detection of specimen contamination in routine histopathology by HLA class II typing using the polymerase chain reaction and sequence specific oligonucleotide probing. J Pathol 173:243–248

Bateman AC, Sage DA, Al-Talib RK, Theaker JM, Jones DB, Howell WM (1996) Investigation of specimen mislabelling in paraffin-embedded tissue using a rapid, allele-specific, PCR-based HLA class II typing method. Histopathology 28:169–174

Bateman AC, Hemmatpour SK, Theaker JM, Howell WM (1997) Nested polymerase chain reaction-based HLA class II typing for the unique identification of formalin-fixed and paraffin-embedded tissue. J Pathol 181:228–234

Bateman AC, Turner SJ, Theaker JM, Howell WM (1998) HLA-DQB1*0303 and *0301 alleles influence susceptibility to and prognosis in cutaneous malignant melanoma in the British Caucasian population. Tissue Antigens 52:67–73

Bein G, Haase D, Schult J, Eirmann TH, Kirchner H (1994) Semiautomated HLA-DQB1 typing by fluorescent dye photometry of amplified DNA on microtiter plates. Hum Immunol 39:1–8

Bettinotti MP, Mitsuishi Y, Bibee K, Lau M, Terasaki PI (1997) Comprehensive method for the typing of HLA-A, B, and C alleles by direct sequencing of PCR products obtained from genomic DNA. J Immunother 20:425–430

Bidwell JL (1992) Applications of the polymerase chain reaction to HLA class II typing. Vox Sang 63:81–89

Bidwell JL, Bidwell EA, Savage DA, Middleton D, Klouda PT, Bradley BA (1988) A DNA-RFLP typing system that positively identifies serologically well-defined and ill-defined HLA-DR and DQ alleles, including DRw10. Transplantation 45:640–646

Blasczyk R, Hahn U, Wehling J, Huhn D, Salama A (1995) A complete subtyping of the HLA-A locus by sequence-specific amplification followed by direct sequencing or single-strand conformation polymorphism analysis. Tissue Antigens 46:86–95

Bodmer JG, Marsh SGE, Albert ED, Bodmer WF, Bontrop RE, Dupont B, Erlich HA, Hansen JA, Mach B, Mayr WR, Parham P, Petersdorf EW, Sasazuki T, Schreuder GMTh, Strominger JL, Svejgaard A, Terasaki PI (1999) Nomenclature for factors of the HLA system, 1998. Eur J Immunogenet 26:81–116

Bodmer WF (1995) Evolution and function of the HLA region. Cancer Surv 22:5–16

Braud VM, Allan DSJ, McMichael AJ (1999) Functions of nonclassical MHC and non-MHC-encoded class I molecules. Curr Opin Immunol 11:100–108

Brodsky FM, Lem L, Bresnahan PA (1996) Antigen processing and presentation. Tissue Antigens 47:464–471

Brown JH, Jardetsky TS, Gorga JC, Stern LJ, Urban RG, Strominger JL, Wiley DC (1993) Three-dimensional structure of the human class II histocompatibility antigen HLA-DR1. Nature 364:33–39

Browning MJ, Krausa P, Bodmer JG (1994) Low resolution HLA-A locus typing by multiplex ARMS-PCR. Hum Immunol 39:134 (abstract)

Bunce M, O'Neill CM, Barnardo MCNM, Krausa P, Browning MJ, Morris PJ, Welsh KI (1995) Phototyping: comprehensive DNA typing for HLA-A, B, C, DRB1, DRB3, DRB4, DRB5 and DQB1 by PCR with 144 primer mixes utilizing sequence-specific primers (PCR-SSP). Tissue Antigens 46:355–367

Clay TM, Bidwell JL, Howard MR, Bradley BA (1991) PCR fingerprinting for selection of HLA matched unrelated marrow donors. Lancet 337:1049–1052

Day IN, Spanakis E, Palamand D, Weavind GP, O'Dell SD (1998) Microplate-array diagonal-gel electrophoresis and melt-MADGE: tools for molecular-genetic epidemiology. Trends Biotechnol 16:287–290

Day IN, O'Dell SD, Spanakis E, Weavind GP (1999) Microplate array diagonal gel electrophoresis (MADGE), CpG-PCR and temporal thermal ramp-MADGE (Melt-MADGE) for single nucleotide analyses in populations. Genet Anal 14:197–204

Dyer PA, Claas FHJ (1997) A future for HLA matching in clinical transplantation. Eur J Immunogenet 24:17–28

Faas SJ, Menon R, Braun ER, Rudert WA, Trucco M (1996) Sequence-specific priming and exonuclease-released fluorescence detection of HLA-DQB1 alleles. Tissue Antigens 48:97–112

Ferencik S, Grosse-Wilde H (1993) A simple photometric detection method for HLA-DRB1 specific PCR-SSP products. Eur J Immunogenet 20:123–125

Fernandez-Vina MA, Bignon JD (1997) Primers and oligonucleotide probes (SSOP) used for DNA typing of HLA class II alleles. In: Charron D (ed) Genetic diversity of HLA – functional and medical implications. Proceedings of the 12th International Histocompatibility Workshop and Conference, vol 1. EDK, Paris, pp 584–595

Giorda R, Lampasona V, Kocova M, Trucco M (1993) Non-radioisotopic typing of human leukocyte antigen class II genes on microplates. Biotechniques 15:918–925

Guttridge MG, Thompson J, Worwood M, Darke C (1998) Rapid detection of genetic mutations associated with haemochromatosis. Vox Sang 75:253–256

Hemmatpour SK, Dunn PPJ, Evans PR, Green A, Howell WM (1998) Functional characterization and exon 2-intron 2-exon 3 gene sequence of HLA-B*2712 as found in a British family. Eur J Immunogenet 25:395–402

Howell WM, Holgate ST (1996) Human leukocyte antigen genes and allergic disease. In: Hall IP (ed) Genetics of asthma and atopy. Karger, Basel, pp 53–70

Howell WM, Navarrete C (1996) The HLA system: an update and relevance to patient-donor matching strategies in clinical transplantation. Vox Sang 71:6–12

Howell WM, Evans PR, Wilson PJ, Spellerberg MB, Smith JL (1989) Comparison of serological, cellular and DNA-RFLP HLA matching in the selection of related bone marrow donors. Bone Marrow Transplant 4:63–68

Howell WM, Leung ST, Jones DB, Nakshabendi I, Hall MA, Lanchbury JS, Ciclitira PJ, Wright DH (1995) HLA-DRB, -DQA and -DQB polymorphism in celiac disease and enteropathy-associated T-cell lymphoma: common features and additional risk factors for malignancy. Hum Immunol 43:29–37

Kennedy LJ, Poulton KV, Thomson W, Williams F, Middleton D, Howell WM, Tarassi K, Papasteriades C, Albert E, Fleischhauer K, Chandanayingyong D, Tiercy JM, Juji T, Tokunaga K, Ollier WER (1997) HLA class I DNA typing using sequence specific oligonucleotide probes. In: Charron D (ed) Genetic diversity of HLA – functional and medical implications. Proceedings of the 12th International Histocompatibility Workshop and Conference, vol 1. EDK, Paris, pp 216–225

Kimura A, Sasazuki T (1992) 11th international histocompatibility wokshop reference protocol for the HLA DNA-typing technique. In: Tsuji K, Aizawa M, Sasazuki T (eds) HLA 1991. Proceedings of the 11th International Histocompatibility Workshop, vol 1. Oxford University Press, Oxford, pp 397–419

Kimura A, Fukuda Y, Hori H, Hoshino S, Sasazuki T, Dohi K (1993) Polymerase chain reaction-single-strand conformation polymorphism analysis of HLA-DP genes and its application in transplantation. Transplant Proc 25:199–202

Krausa P, Browning M (1996) Detection of HLA gene polymorphism. In: Browning M, McMichael A (eds) HLA and MHC: genes, molecules and function. BIOS Scientific Publishers Limited, Oxford, pp 113–137

Lalvani A, Hill AVS (1998) Cytotoxic T-lymphocytes against malaria and tuberculosis: from natural immunity to vaccine design. Clin Sci 95:531–538

Lanier LL (1998) Follow the leader: NK cell receptors for classical and non-classical MHC class I. Cell 92:705–707

Lienert K, Russ G, Lester S, Bennett G, Gao X, McCluskey J (1996) Stable inheritance of an HLA-"blank" phenotype associated with a structural mutation in the HLA-A*0301 gene. Tissue Antigens 48:187–191

Luedeck H, Blasczyk R (1997) Fluorotyping of HLA-C: differential detection of amplicons by sequence-specific priming and fluorogenic probing. Tissue Antigens 50:627–638

Madrigal JA, Arguello R, Scott I, Avakian H (1997a) Molecular histocompatibility typing in unrelated donor bone marrow transplantation. Blood Rev 11:105–117

Madrigal JA, Scott I, Arguello R, Szydlo R, Little A-M, Goldman JM (1997b) Factors influencing the outcome of bone marrow transplants using unrelated donors. Immunol Rev 15:153–166

Metcalfe P, Waters AH (1993) HPA-1 typing by PCR amplification with sequence-specific primers (PCR-SSP): a rapid and simple technique. Br J Haematol 85:227–229

Miller SA, Dykes DD, Polesky HF (1988) A simple salting out procedure for extracting DNA from human nucleated cells. Nucleic Acids Res 16:1215

Mittal KK, Mickey MR, Singal DP, Terasaki PI (1968) Serotyping for homotransplantation: refinement of microdroplet lymphocyte cytotoxicity test. Transplantation 6:913–927

Mytilineos J, Scherer S, Opelz G (1990) Comparison of RFLP-DR beta and serological HLA-DR typing in 1500 individuals. Transplantation 50:870–873

Newton CR, Graham A, Heptinstall LE, Powell SJ, Summers C, Kalsheker N, Smith JC, Markham AF (1989) Analysis of any point mutation in HLA. The amplification refractory mutation system (ARMS). Nucleic Acids Res 17:2503–2516

Olerup O, Zetterquist H (1992) HLA-DR typing by PCR amplification with sequence-specific primers (PCR-SSP) in 2 hours: an alternative to serological DR typing in clinical practice including donor-recipient matching in cadaveric transplantation. Tissue Antigens 39:225–235

Opelz G, Mytilineos J, Scherer S, Dunckley H, Trejaut J, Chapman J, Fischer G, Fae I, Middleton D, Savage D, Bignon J-D, Bensa J-C, Noreen H, Albert E, Albrecht G, Schwarz V (1993) Analysis of HLA-DR matching in DNA-typed cadaver kidney transplants. Transplantation 55: 782–785

Parham P (1992) Typing for HLA class I polymorphism: past, present and future. Eur J Immunogenet 19:347–359

Perrey C, Turner SJ, Pravica V, Howell WM, Hutchinson IV (1999) ARMS-PCR methodologies to determine IL-10, TNF-a, TNF-β and TGF-β1 polymorphisms. Transplant Immunol 7:127–128

Powis SH, Teisserenc H (1997) TAP, LMP and HLA-DM typing protocols. In: Charron D (ed) Genetic diversity of HLA – functional and medical implications. Proceedings of the 12th International Histocompatibility Workshop and Conference, vol 1. EDK, Paris, pp 226–235

Procter J, Crawford J, Bunce M, Welsh KI (1997) A rapid molecular method (polymerase chain reaction with sequence-specific primers) to genotype for ABO blood group and secretor status and its potential for organ transplants. Tissue Antigens 50:475–483

Ramon DS, Arguello JR, Cox ST, McWhinnie A, Little A-M, Marsh SGE, Madrigal JA (1998) Application of RSCA for the typing of HLA-DPB1. Hum Immunol 59:734–747

Savage DA, Tang JP, Wood NA, Evans J, Bidwell JL, Wee JL, Oei AA, Hui KM (1996) A rapid HLA-DRB1*04 subtyping method using PCR and heteroduplex generators. Tissue Antigens 47: 284–292

Shintaku S, Fukuda Y, Kimura A, Hoshino S, Tashiro H, Sasazuki T, Dohi K (1993) DNA conformation polymorphism analysis of DR52 associated HLA-DR antigens by polymerase chain reaction: a simple, economical and rapid examination for HLA matching in transplantation. Jpn J Med Sci Biol 46:165–181

Singal DP, Ye M (1998) HLA-DM polymorphisms in patients with rheumatoid arthritis. J Rheumatol 25:1295–1298

Spielman RS, McGinnis RE, Ewens WJ (1993) Transmission test for linkage disequilibrium: the insulin gene region and insulin dependent diabetes mellitus (IDDM). Am J Hum Genet 52: 506–516

Stern LJ, Wiley DC (1994) Antigenic peptide binding by class I and class II histocompatibility proteins. Structure 2:245–251

Tatari Z, Fortier C, Bobrynina V, Loiseau P, Charron D, Raffoux C (1995) HLA-Cw analysis by PCR-restriction fragment length polymorphism: study of known and additional alleles. Proc Natl Acad Sci USA 92:8803–8807

Thorsby E (1997) Invited anniversary review: HLA and disease. Hum Immunol 53:1–11

Thorsby E, Ronningen KS (1992) Role of HLA genes in predisposition to develop insulin-dependent diabetes mellitus. Ann Med 24:523–531

Trowsdale J (1995) "Both man & bird & beast": comparative organization of MHC genes. Immuno-
     genetics 41:1–17
Tuokko J, Pushnova E, Yli-Kerttula U, Toivanen A, Ilonen J (1998) TAP2 alleles in inflammatory
     arthritis. Scand J Rheumatol 27:225–229
Uryu N, Maeda M, Ota M, Tsuji K, Inoko H (1990) A simple and rapid method for HLA-DRB and
     -DQB typing by digestion of PCR-amplified DNA with allele specific restriction endonucle-
     ases. Tissue Antigens 35:20–31
Van Ham SM, Tijn EPM, Lillemeir BF, Gruneberg U, van Meijgaarden KE, Pastoors L, Verwoerd
     D, Tulp A, Canas B, Rahman D, Ottenhof THM, Pappin DJC, Trowsdale J, Neefjes J (1997) HLA-
     DO is a negative modulator of HLA-DM-mediated MHC class II peptide loading. Curr Biol
     7:950–957
Vogt AB, Kropshofer H, Hammerling GJ (1997) How HLA-DM affects the peptide repertoire
     bound to HLA-DR molecules. Hum Immunol 54:170–179

7 Microplate Array Diagonal Gel Electrophoresis (MADGE) Methodologies: The First Five Years

IAN N.M. DAY, EMMANUEL SPANAKIS, LESLEY J. HINKS, ANCA VOROPANOV, XAIO-HE CHEN, and SANDRA D. O'DELL

7.1 Introduction

Recently, the Wellcome Trust (UK) and a consortium of major pharmaceutical companies have announced a 2-year initiative costing £28 million to identify 300,000 and map 150,000 human single nucleotide polymorphisms, information which will be available in a public domain database. This is driven by the recognition that systematic disease association studies are becoming feasible and will benefit many areas of future health care developments. We are concerned with such molecular genetic epidemiology and, in 1994, described a system, "MADGE" (microplate array diagonal gel electrophoresis), which has a number of advantages for laboratory molecular genetic analysis of population samples (Day and Humphries 1994a,b). This system contests the traditional wisdom that electrophoresis should be avoided and liquid-phase analyses developed, where high-throughput applications are contemplated. The system combines liquid-phase microplate compatibility (laboratory transfers and informatics), convenient setup and usage of polyacrylamide gel (PAGE) (which has higher resolution than agarose post-PCR). Short track length/short run time, compactness and scalability with hand-sized robust slab gels accessible for direct human interaction, i.e. mini- not microscale, are additional features. Start-up costs are minimal, so "Third World" laboratories can also apply the system. Essentially, any analysis that can be reduced to short tracks (e.g. <3 cm) gains the throughput advantages. One technician can process 10–100 gels (1000–30,000 tracks) per day according to the application and the MADGE implementation in use. Our principal focus has been on diversity of categories of analysis of polymorphic PCR amplicons in the context of human genetic variation and polygenic disease traits. However, MADGE is applicable to any other species and potentially to many other categories of biomolecule (protein etc), and we can only speculate on future utilisation. Genome mapping (simple but very high-throughput PCR checking gels) has already emerged (Watanabe et al. 1999). We restrict our discussions here to those pertinent to molecular genetic epidemiology.

In 1807, Ferdinand Frederic Reuss observed under a microscope the migration of colloidal particles in an electric field, perhaps the first electrophoretic

Principles and Practice
Molecular Genetic Epidemiology – A Laboratory Perspective
Ian N.M. Day (Ed.)
© Springer-Verlag Berlin Heidelberg 2002

separation. During the early 1970s, both agarose and polyacrylamide electrophoresis gel usage evolved to slab gel formats (Studier 1973; Sugden et al.
1975) much as they are still found in the laboratory today. Throughput requirements were not an issue at that time. Indeed, slabs were an improvement over
tube gels. For both tubes and modern capillaries, each "lane" is separate from
any other and cross-referencing requires reliance either on an external frame
of reference (e.g. time of elution) or an internal frame (e.g. co-electrophoresed
mobility standards). Slabs give convenient alignment of tracks. MADGE, in the
process of achieving microplate compatibility, uses 96 (higher density in newer
formats) track origins (wells) with 8×12 array locations at the 9-mm pitch
identical to microplates devised 30 years ago by Dynatech and now a highly
established industry standard for liquid-phase operations. The long axis of the
array is at an acute angle, e.g. $18.4°$, relative to the direction of electrophoresis
(e.g. Fig. 7.1b) so that a track length of 25–30 mm is available. It should be noted
that each track is specifically isolated (like a set of tube gels or capillaries) but
the tracks are in a slab. The two-piece kit, i.e. dimensional gel former and piece
of backing glass or plastic (Fig. 7.1) for the gel, both excludes air and gives
a perfectly flat open face to the resultant gel, simplifying the production of
agarose, PAGE, and other gel matrix or composite gels. The combination of
features in the MADGE gel design gives substantial advantages (as described
above and below), but the features of this system can be used in isolation, for
example, the simplified production of open-faced horizontal PAGE gels (H-
PAGE; Day and Humphries 1994a) but with standard row(s) of wells.

Molecular genetic epidemiological studies demand economy, sample parallelism, convenience of setup and accessibility to small as well as large laboratories, goals inadequately met by existent approaches. More systematic
association studies will demand the analysis of many thousands of candidate
variations, commonly single nucleotide variations. Power studies (see Risch
and Merikangas 1996) indicate that for common disease traits where individual genotypes make smaller percentage contributions, cohorts, trios or other
architectures comprising hundreds or thousands of individuals or nuclear
families will be essential. Such studies can be prohibitively expensive both in
phenotyping and genotyping, and, therefore, the major hurdles to overcome
are in cost reduction. These needs have driven the laboratory technology
inventions described here. Our philosophy has been that electrophoresis still
has many advantages over homogeneous systems, including ready analysis of
parameters such as size, shape and charge of molecular moieties. It is also well
known and available to all users, in contrast to the expensive hardware implicit
in homogeneous techniques, and a range of well-established methodologies
are already centred around electrophoresis (ARMS, restriction analysis,
heteroduplex analysis, DGGE, SSCP, protein electrophoresis, etc.). However,
gel preparation, long track lengths, incompatibility with industry standard
microplates, vertical format polyacrylamide gels, and a host of other general
inconveniences, make electrophoresis an unattractive (although used) option
to underpin laboratory studies of genetic diversity within populations. We have

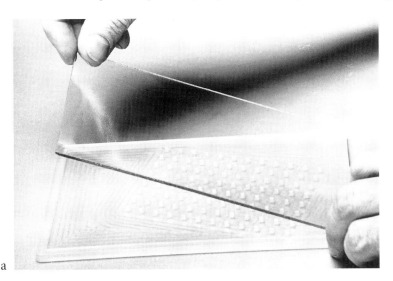

a

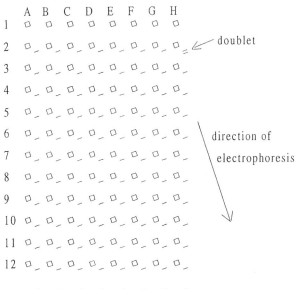

b

Fig. 7.1a,b. Microplate array diagonal gel electrophoresis (MADGE). **a** Preparation of a MADGE gel, original and simplest format. The plastic former is laid horizontal with teeth upward. Acrylamide gel mix is poured into the 'swimming pool' which contains the teeth. A Sticky-Silane coated glass plate is laid over. Once the gel has set (1–5 min if higher concentrations of polymerisation agents are used), the glass plate is prised off, bearing an open-faced microplate-compatible 96-well PAGE gel suitable for submersible or semi-dry use. **b** Schematic of a MADGE gel image. The *squares* represent the wells: using ethidium bromide imaging, as in Fig. 7.3, the wells are only appreciable as a dark image on a dark background, whereas the bands stand out brightly in a regular array. The *lines* represent bands in the lanes. Although seemingly complex, the human eye is efficient at pattern recognition and can readily identify different patterns. For example, here, only track H1 contains a doublet. The setup is such that well A1 is placed nearest the cathode

Table 7.1. History of MADGE methodologies

1993	Concept and patent application
1994	Primary publication (Day and Humphries 1994a)
1995	Applications to checking, sizing ($\geq$10% mobility differences) and RFLPs (Bolla et al. 1995; O'Dell et al. 1995)
1996	CpG-PCR. PCR reconfigurations to very short PCRs enabled by convenient access to PAGE using MADGE (O'Dell et al. 1996)
1997	MADGE software availability (Phoretix Int.; http://www. phoretix.com) First de novo SNP/complex trait association discovery using MADGE (IGF2 ApaI site/population obesity) (O'Dell et al. 1997)
1998	Key review (Day et al. 1998), presentations at HUGO sponsored meetings – First International SNP Meeting, Skokloster, Sweden and Mutation Detection Workshop, Sanger Centre, Cambridge, UK Brief descriptions of melt-MADGE for de novo mutations scanning (1000 amplicons per hour) Commercial products (MadgeBio Ltd.; http://www.madgebio.freeserve.co.uk) At least 50 laboratories worldwide known to be using MADGE, including one or two with very high-throughput applications, one or two for proteins, peptides and enzyme electrophoresis
1999	Application to construction of a 5255-marker radiation hybrid map of the rat genome (Watanabe et al. 1999) In the pipeline: 192-well (ARMS-MADGE) and 384-well MADGE; larger 'stand alone' multiple gel systems; application to VNTRs and microsatellite polymorphisms (<2% mobility differences); approaches to multiple gel re-use Total PCR-MADGE tracks known to be well into the millions.
2000+	? Widening usage ? extensive protein applications ? more extensive commercialisation ? Complete automation ? quantitative applications ? Preparative applications ? combination with 'intelligent' gel matrices (e.g. affinity oligos)

aimed to eliminate these disadvantages. Several representative developments (Table 7.1) are described here.

1. Microplate array diagonal gel electrophoresis (MADGE), for PCR checking and for assays of known single nucleotide variants (Day and Humphries 1994a,b; Bolla et al. 1995; Day et al. 1995a,b; O'Dell et al. 1995).
2. CpG-PCR, a final common analytical pathway, compatible with MADGE, for examining any selected CpG site (O'Dell et al. 1996).
3. Melt-MADGE, for rapid, high-throughput, de novo mutation scanning of PCR products (Day et al. 1995b, 1998; Spanakis et al., submitted).
4. Software for MADGE gel image analysis.
5. A brief overview is given of further developments which are completed and are in preparation as full original papers: 192- and 384-well MADGE for ARMS, heteroduplex generator and higher-throughput applications; and

higher resolution MADGE, such as for minisatellite and tetranucleotide microsatellite polymorphisms.

7.2 Microplate Array Diagonal Gel Electrophoresis (MADGE)

Electrophoresis of DNA has been performed traditionally either in an agarose or a polyacrylamide gel matrix. Much effort has been directed to produce improved quality agaroses capable of high resolution, but for small fragments, such as those from PCR and post-PCR digests, polyacrylamide still offers the highest resolution. Although agarose gels can easily be prepared in an open-faced format to gain the conveniences of horizontal electrophoresis, acrylamide does not polymerise in the presence of air and the usual configurations for gel preparation lead to electrophoresis in the vertical dimension.

The original MADGE format (Day and Humphries 1994a) uses a two-dimensional plastic former in conjunction with one glass plate coated with γ-(methacryloxy)propyltrimethoxysilane (Sticky Silane). The plastic former contains a 2 mm deep, 100×150 mm rectangular "swimming pool". Within the pool, there are 96 2-mm^3 "teeth" (well formers) in an 8×12 array with 9-mm pitch directly compatible with 96-well plates. The array is set on a diagonal of 71.6° relative to the long side of the "pool", which is parallel to the eventual direction of electrophoresis, giving gel track lengths of 26.5 mm. The gel-former is placed horizontally, the acrylamide mix is poured into the pool and the glass plate overlaid (Fig. 7.1a). After the gel has set, the glass plate is prised off, bearing its open-faced 96-well gel. A full description of the apparatus and gel setup can be found in the original report of Day and Humphries (1994a). It should be noted that, although the system may sound complex, many users have noted that this is the simplest gel system they have ever used. Additionally, the human eye/brain is extremely good at pattern recognition and the small-scale user, once accustomed, is not obliged to use image analysis software. A 26.5-mm track length (2×4 diagonal of array), such as that in Fig. 7.1b, or greater, such as the 2×6 diagonal in the molecular weight markers image in Fig. 7.2, using a polyacrylamide gel matrix, is efficient for many post-PCR analyses involving simple band pattern recognition. Since the simplest situation for post-PCR analysis is to load PCR product directly onto the gel, the question arises about the effects of carry-over buffers and salts in the sample solutions on separation in MADGE. Our typical PCRs contain 50 mM KCl and loading 5 µl leads to a reasonable resolution of patterns where relative mobility differences between bands are 5% or greater.

Extremely high salt digests or other less typical solutions may occasionally present a problem. Rather than resort to laborious ethanol precipitations of many microplates of amplicons, or other expensive purification procedures, we have usually been able to "dilute away" the solute problem by diluting the sample in the gel buffer without any compromise to band detection. If ethid-

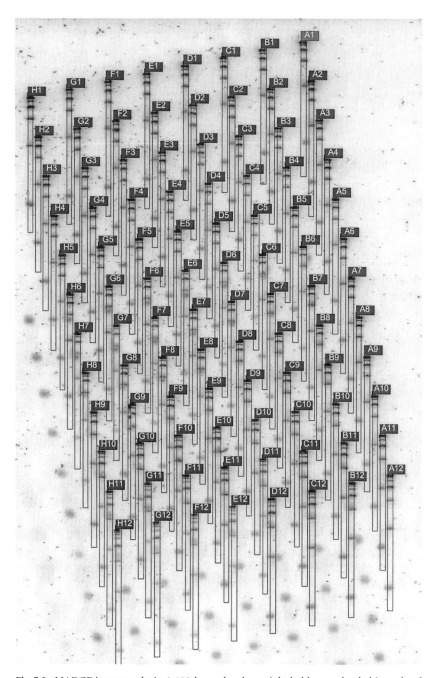

Fig. 7.2. MADGE image analysis. A 100-bp molecular weight ladder was loaded in each gel track and electrophoresis undertaken. The digital image of the ethidium bromide stained gel is shown here undergoing the first step of computer analysis involving gel track identification. The gel tracks can then be subjected to lateral alignment, band identification, quantitation, Rf analysis, data export, etc.

ium bromide turns out then to provide insufficient sensitivity, order of magnitude improvements in sensitivity are readily achievable with newer intercalating dyes such as Vistra Green. Indeed, poorer resolution of doublets with less than 5% mobility differences has often simply been the consequence of band overloading. The thicker (2-mm) gel used with transverse imaging would also offer lower resolution if bands were significantly tilted in orientation from the glass toward the gel surface. This is no greater challenge than for horizontal submerged agarose gels and generally this has not been a problem at all in our laboratory. Attention to detail of buffers and ions in both gel and sample (Kozulic 1994) may improve results further but for the typical applications described it must be stressed that direct loading of samples usually gives perfectly readable results. Gels are robust, reusable, directly stackable for storage or use, and fully compatible with industry-standard PCR microplates, thus minimising procedural information transfers and enabling direct 96-channel pipetting. However, considerable advantage is gained even with 8- and 12-channel pipettes. Use of passive replicators may also be possible and the horizontal gel format will simplify fully robotic handling. Kits, ready-made gels, other components and MADGE image analysis software are commercially available from several sources. One individual can conveniently handle up to 20 gels (~2000 tracks) per hour.

MADGE can be arranged in various formats, for example, free-floating gels or PAGE gels adherent to a single thin float-glass plate. These different formats have been reviewed recently (Day et al. 1998). For Melt-MADGE (see below), the glass backing provides a convenient support for shelving gels in a stack and also provides an efficient medium for thermal transfer in conjunction with Melt-MADGE. Although we do not currently know the worldwide patterns of MADGE usage, we know of millions of sample-PCR analyses. The throughput has enabled us to relax the stringency of selection of "candidate" single nucleotide polymorphisms for association studies of genes in human obesity. Using this approach, preliminary analyses of several sites in the insulin-like growth factors/receptors gene family were undertaken. A positive association in the *IGF2* gene was identified and rapidly replicated (O'Dell et al. 1997) and a complete scan of this region is currently underway. The initial candidate screen benefited from a combination of ease of setup (RFLPs can be more readily set up for analysis by PCR-MADGE than by ASOs), throughput, low cost and ease of data calling.

7.3 CpG-PCR

CpG sites are frequently methylated in the mammalian genome. This epigenetic phenomenon leads to mutational instability, and also plays a role in imprinting. Additionally it is a correlate of developmental and carcinogenic changes, and could conceivably be implicated in transgenerational program-

CpG-PCR

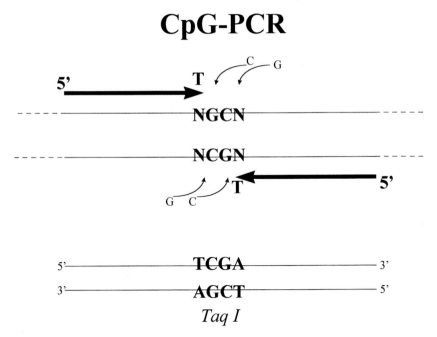

Fig. 7.3. Schematic illustration of CpG-PCR. Both PCR primers bear 3′ T bases (natural or mismatched). In the presence of an intact CpG site, a 5′-TCGA *TaqI* restriction site results, enabling a simple assay to screen for mutation at any CpG site, or to genotype single nucleotide polymorphism residing at a CpG site. The digested halves of the PCR product are of the same size and therefore co-run, but have a much greater mobility than the uncut product: pattern resolution is therefore simple, and bands readily detected. The short fragments, e.g. approx. 50 and 25 bp, are fully compatible with polyacrylamide MADGE

ming. It is estimated that 10–20% of all human single nucleotide variation (polymorphism and rare mutations) resides at CpG sites, and therefore a general approach will be valuable (Cooper and Krawczak 1990).

A final common pathway has been devised (Fig. 7.3) for the analysis of de novo mutation at any CpG site. Artificial restriction sites can be introduced in known DNA sequences by using either or *both* sense and antisense mismatched PCR primers. Siting of the primers directly adjacent 5′ and 3′ to the CpG site yields a 52-bp PCR product, resulting from the sum of the two 25mer oligonucleotides plus the two intervening bases (C and G), and also yields consistent digestion fragments. Three out of four possible four-base palindromes (*TaqI*, *HhaI* and *MspI*) were investigated for mutations R329X and E80K in the human LDL receptor gene, and for mutations R395W and R612C, the *TaqI* site was forced using PCR in which *both* primers had 3′ mismatched T. Both empirically and on theoretical grounds, *TaqI* is the forced restriction site of choice, in which both PCR primers have a 3′ T base. The approach has been adapted to the high-throughput microplate diagonal gel electrophoresis (MADGE) system for analysis of any selected CpG site.

CpG-PCR also illustrates that polyacrylamide-MADGE readily permits analysis of "ultrashort" PCRs. This has major implications for PCR planning. Many laboratories and experiments use unnecessarily long PCRs, because shorter fragments cannot conveniently be analysed or resolved using standard horizontal submersible agarose gels. Excess high molarity sequence beyond that necessary for the analysis is undesirable, as it may compromise the PCR: this is particularly notable in multiplex PCR. Furthermore, shorter products show greater mobility shifts on restriction digestion, simplifying pattern resolution and recognition (see above).

7.4 Temporal Thermal Ramp MADGE Electrophoresis (Melt-MADGE)

Since MADGE implies a complex spatial configuration of sample wells and tracks, methods involving spatial gradients will be difficult to establish. Our first efforts, however, were focused on the objective of simplifying the use of allele-specific oligonucleotides by determining complete melting profiles rather than single temperature "snapshots" using real time thermal ramp electrophoresis (Day et al. 1995b). At that time we had not explicitly recognised that real-time-variable-temperature electrophoresis, i.e. temporal thermal ramps, would in general be the ideal substitute for spatial chemical gradients for MADGE, for applications involving duplex melting: however, this is now clear. This generality has therefore been termed Melt-MADGE, for which we have built a prototype apparatus. Two applications have been developed, firstly for the determination of complete ASO melting profiles in which the "freed" rather than the "bound" ASO is examined (Day et al. 1995b) and, secondly, for de novo mutation scanning of complete duplexes, effectively a reconfiguration of denaturing gradient gel electrophoresis (DGGE) in which the temporal thermal ramp(s) replaces the spatial chemical denaturant gradient as the mechanism of interrogation of full duplex base-pairing. There has recently been a profusion of DGGE variations in the literature, using temperature in space (temp gradient gel electrophoresis, TGGE; Henco et al. 1994) and temperature in time (temporal thermal gradient electrophoresis, TTGE) for capillary electrophoresis (Gelfi et al. 1996): these remain inconvenient (generally large apparatus), have an inherently low throughput (one row of samples across the top of the gel), and are expensive (particularly multi-capillary instrumentation). Our system, designated Melt-MADGE, enables analysis of almost 1000 PCR products with 1-h run times in a 2-l tank, using ten palm-sized MADGE gels. Additional methods under development also utilise spatial gradients in conjunction with MADGE and horizontal gels (Spanakis and Day, unpubl.).

Nomenclature has become an issue. The term "gradient" should be reserved for space-dependence, "ramp" for time-dependence (L.S. Lerman, pers. comm.). We note, furthermore, that a "denaturing gradient" is not necessarily chemical; it can be thermal. Hence TGGE is a variant of DGGE, TTGE is a misnomer. Our gels are horizontal, not vertical. Samples can be loaded either transversely or parallel to the direction of electrophoresis. Mnemonic nomenclature has been adopted in-house, but if systematic nomenclature can be derived, we favour the core term "melt".

The apparatus used is purpose-built (Day et al. 1995b). In brief, a $10 \times 10 \times 15$ cm electrophoresis tank contains buffer which is continuously mixed to ensure spatial thermal homogeneity, and the temperature is varied real-time by programmable software controlling heating and cooling systems and receiving feedback from temperature sensors in the tank. Overall, this is a little more complex than a PCR machine, although, for PCR, low thermal mass for rapid temperature shifts is the objective, whereas, for Melt-MADGE, the aim is high thermal mass for stability of small temperature shifts. The initial prototype accommodates 10–12 horizontal MADGE gels spaced in a carrier stack. Each 2-mm-thick gel is adherent to and supported by a 2-mm glass plate and is also covered by 2-mm glass, with spacing between each glass/gel/glass sandwich for buffer circulation. Since buffer and hence thermal circulation is vigorous, and the glass is both a good thermal conductor and also quite thin, and the thermal ramps are relatively slow and shallow, reasonable spatial thermal homogeneity is achieved, at least to within 0.1°C, as determined by a reference platinum resistance thermometer.

De novo mutation detection is also an important component of research in genetic disease. For mutations not known in advance, the choice at present is between direct sequencing, de novo scanning techniques (Cotton 1997) to target sequencing more efficiently to appropriate regions, and "chip" technologies, equivalent to direct resequencing by multiple parallel oligonucleotide hybridisation (e.g. Chee et al. 1996). For the identification of rare variants, important criteria are the degree of parallelism of samples possible, the number of bases analysed per "run", and the total number of bases an individual can scan per day. For heterozygote sequencing in humans, the difficulties of accurate base calling (essential for every base to avoid excessive numbers of false positive calls) substantially restrict sequencing as a first-line approach (Parker et al. 1996). Chip technologies (Goffeau 1997), while promising, have low capability for sample parallelism; a chip is expensive ($5000–10,000) for any selected sequence and also requires an optimisation process; and the approach will demand high capital expenditure on chip-reading hardware.

Scanning technologies are much used, but intense activity to develop improved approaches reflects the importance and shortfalls in this field (Cotton 1997). Such technologies generally rely on electrophoresis and give some information about the mutation from the resultant band pattern. SSCP appears to be favoured by research laboratories; it is easy to implement, as evidenced by 2000 or so Medline references, although sensitivity to all base changes may range from only 30 up to 90% (Sheffield et al. 1993). However, the average sensitivity of SSCP to base changes remains uncertain and the approach in the cited analysis has itself been questioned (Hayashi and Yandell 1993). Denaturing gradient gel electrophoresis (DGGE) is favoured by diagnostics laboratories: it is laborious to set up but generally more sensitive and robust (Moyret et al. 1994).

One objection to DGGE raised by some investigators is the additional cost of oligo clamp synthesis and the occasional adverse influence of the clamp on the PCR reaction itself. The additional cost element corresponds to one or two additional oligos, or to a fluorescence end label, which is not necessary in MADGE applications. It is small compared with costs of sequencing many templates, and seems a reasonable compromise to achieve a "PCR and run" technique evading other intermediate steps and costs. It has not been our experience that GC clamps compromise many PCRs. We have previously reported modifications of SSCP to achieve scanning rates of approximately 20,000 bases per day per person (Whittall et al. 1995; Day et al. 1997). However, SSCP has limitations related to the unpredictable effect (if any) of a base change, with resulting inability to predict, control and maximise band resolution, as well as the relative inconvenience of detection systems suitable for single-stranded DNA. The folding and tertiary structure, and hence electrophoretic mobility, of a single strand depend on chance internal base pairing and other influences of bases on overall tertiary structure. The effectiveness is therefore a matter of chance with which a particular single-strand sequence will interrogate all its own internal bases. By contrast, in DGGE the base sequence of one strand is being interrogated in a predictable fashion by the complementary strand.

The principle of denaturing gradient gel electrophoresis (DGGE; Fischer and Lerman 1983) is that strands with a sequence difference or heteroduplex molecules melt and alter their electrophoretic mobility under slightly different conditions of denaturation encountered at some point in space. This is achieved using a gel with either a chemical or thermal gradient from the row of wells of origin down the length of the tracks. However, the use of a spatial gradient has several disadvantages. Firstly, preparing gradient gels is cumbersome and inflexible and sophisticated (non-linear) gradients are practically impossible to prepare. Secondly, the spatial gradient introduces the constraint of a single, rather than multiple rows or arrays of wells, which severely limits throughput. Finally, the setup does not readily allow the use of very short track lengths, which further reduces the possibility of utilising high-density arrays of tracks. Melt-MADGE uses the same principle as DGGE or TGGE but **exchanges the dimension of the gradient, so that it is a ramp in time rather than a gradient in space.** Thus the PCR products for analysis are loaded on a homogeneous gel and the temperature of the entire gel is raised during the course of the electrophoresis. DNA fragments which form a melted domain at a lower temperature (e.g. heteroduplexes) will display reduced mobility at an earlier stage of the run, and hence will not migrate as far as their more thermostable cognate homoduplexes. A Melt-MADGE proof-of-principle experiment is shown in Fig. 7.4. The reconfiguration of the denaturing gradient to have temperature as the dependent variable and time as the independent (controlled) variable opens a range of important advantages:

1. It creates the option for high-density arraying of wells and tracks (e.g. MADGE arrays), hence enabling high throughput of analyses.
2. Programmability of the gradient, which can be absolutely arbitrary and use temperature rather than chemicals, is achievable easily in time. Arbitrary gradients (rather than simple linear gradients) would be very difficult in

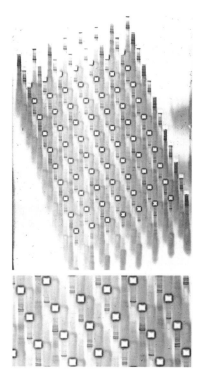

Fig. 7.4. Image of a Melt-MADGE analysis. Using a purpose-built temporal thermal ramp electrophoresis apparatus for "Melt-MADGE", a DGGE-like analysis of GC-clamped PCR products was undertaken. Analysis of *APOE* genotypes. There are three alleles and hence six genotypes, but knowledge of the base changes was not needed in the analysis here: the duplex melting analysis is in effect a de novo mutation scanning analysis and, amongst samples, six different patterns are observed. Homozygotes and heterozygotes for the common alleles e2, e3 and e4 were loaded alternately at various positions of a Melt-MADGE (actual size and detail shown) to show that the gel offers practically the same resolution throughout its surface. Wells are as in a microplate array: rows A, B, . . ., H are from *right* to *left* and columns 1, 2, . . ., 12 from *top* (cathode) to *bottom* (anode). Columns 1, 3, 5, 7, 9 and 11: *A* e3/e4; *B* e2/e4; *C* e3/e3; *D* e2/e3; *E* e3/e4; *F* e4/e4; *G* e3/e3; *H* e2/e3. Columns 2, 4, 6, 8, 10 and 12: *A* e2/e3; *B* e3/e3; *C* e4/e4; *D* e3/e4; *E* e2/e3; *F* e3/e3; *G* e2/e4; *H* e3/e4

space. Gradients as an integral feature of the gel demand special gels. Programmability gives flexibility and convenience.

3. Potential for arbitrary gradients enables very short track lengths, and short run time (e.g. 1h) further enables high track density and hence throughput.

4. The use of ordered arrays (e.g. microplate format of MADGE) much simplifies data handling.

5. Arbitrary programming enables arbitrary resolution, thus reducing the analysis to simple pattern recognition rather than detailed relative mobility measurements. Pattern recognition evades the need for closely juxtaposed tracks or for internal markers. By contrast, the resolution of SSCP is often small, cannot be programmed and hence cannot be arbitrarily great. Thus advantages I(1)–(5) cannot be realised with SSCP.

6. For PCR product analysis for sequence variations (known or de novo), the system will usually enable the user to **"PCR and run"**. This concept represents an operational standard that can only be surpassed by evading electrophoresis altogether, and then only if the liquid-phase procedures are simpler, more economical or more sensitive in some way.

7. Whereas DGGE causes amplicons of similar %(G+C) to migrate to similar positions in the track irrespective of amplicon size, this will not occur in the real time approach, in which migration distance depends heavily on both size and base content. This may facilitate one-dimensional analysis of some amplicon multiplexes more readily.

Melt-MADGE has a resolution theoretically greater than DGGE, since gradients of any complexity can be planned, including, for example, steep initial gradients to resolve heteroduplexes, followed by shallow subsequent gradients to enhance resolution of homoduplexes. Discontinuous gradients to combine analysis of more than one PCR product per run, where their T_m values are quite different, will also be feasible. Sequencing can be used as first-line analysis. However, parallel sequencing of thousands of samples simultaneously would still be a very substantial and expensive task, and, to be successful, no base call in any sequencing reaction should leave any (false) possibility that there could be heterozygosity at that position. In practice, the latter is difficult to achieve (Parker et al. 1996). By contrast, using one strand to interrogate another in a melting assay is an accurate, robust and simple test of bases for perfect complementarity (Guldberg and Guttler 1993). The feasible track usage rate using the first Melt-MADGE prototype is 1000 per 2h, which would correspond to 20–80 manual or automated instrumentation sequencing gels depending on whether a one track/four label or one track/one label sequencing method were employed. Sample parallelism is also 100–1000-fold greater than with a chip, in which one PCR product can be analysed every 15 min. Additionally, chip techniques require post-PCR procedures such as in vitro transcription and fragmentation prior to chip hybridisation (Hacia et al. 1996), rather than being PCR-and-run. Sample parallelism is also problematic with capillary electrophoresis instrumentation, although possible at high equipment cost (Clark and Mathies 1993). The design of a "Melt-MADGE machine" also means that its final cost is likely to be much less than a chip and "chip reader": the former requires electrophoresis in equipment otherwise akin to a PCR thermal cycler, the latter requires photolithography, oligonucleotide synthesiser and purpose-built confocal microscope.

These considerations suggest to us that for molecular genetic epidemiological research, MADGE and Melt-MADGE may better meet the criteria of economy, sample parallelism, convenience of setup and accessibility to large and small laboratories alike, and will thus find an important role in future genetic research in complex diseases.

7.5 Software Applications (Applies to All MADGE Not Just Melt-MADGE)

Image analysis software is often advantageous, as is true for electrophoresis in general. Mobility measurements, band intensities/peak heights, pattern matching, image adjustments, etc., are all feasible and can be systematised to a state of semi-automation. Two features of MADGE to be noted are:

1. Once the grid of lane origins and lane locations have been identified by the software (e.g. Fig. 7.2), other analytical procedures on the lanes are the same as for any other slab gel
2. There are very many lanes, often with a small number of families of patterns in the lanes. "One button click" user identification/validation/review of the "call" for each lane is important for genotyping, as is "templating" the images such that the act of lane finding by the software is essentially automatic for each new image. These requirements may be unique to MADGE and its applications.

7.5.1 Analysing MADGE Gels with Phoretix 1D MADGE

7.5.1.1 Mode

Phoretix (http://www.phoretix.com) 1D software has three main modes of analysis, one of which is for MADGE gels. Selecting MADGE mode changes the appropriate dialogues and displays such that the specific requirements for analysing MADGE gels are incorporated or replace the features and facilities for standard gel analysis. This means that all the densitometry, molecular sizing, Rf adjustments and band pattern patching facilities are available.

7.5.1.2 Identifying Lanes

Any matrix of x rows and y columns can be identified along with desired lane width. A simple click-and-drag procedure then fits this matrix to align with the wells. Adjusting the corners of the matrix overcomes any problems associated with gel distortions or poor imaging. A single lane is added anywhere on the gel by click-and-drag drawing. Once drawn the lane is automatically copied to all other positions using the same angle, length and distance from the well (Fig. 7.2). If necessary, individual lanes may be adjusted for position, angle, width and distortions. All lanes can be identified by number, or a letter number combination, to keep in line with the original microplate. On good gels, identifying lanes takes no more than a few seconds after the default parameters have been set.

7.5.1.3 Background Subtraction

Although many MADGE applications do not require precise quantitation, it is advisable to perform background subtraction simply because it can affect band detection since it changes the significance of changes in individual data points. This can be set as an automatic component of analysis in the "Analysis Template".

7.5.1.4 Band Detection

At this point the gel image changes so that all the lanes are aligned next to each other. Three band detection parameters allow changes to sensitivity, smoothing out noise during band detection and elimination of faint bands. If required, band editing can be performed with simple mouse clicks on the image or on the data profile graphics.

7.5.1.5 Lane Calling

For many MADGE applications, the simplest route to identifying the nature of a lane is to make observational decisions. For this reason the system of "Lane Calling" was developed, which allows the use of the computer's numeric keypad to assign names or codes to each lane.

7.5.1.6 Template Automation

All the analysis procedures can be saved as an "Analysis Template", which means that dragging a Template onto a gel automates or semi-automates future analyses.

7.5.1.7 Flexibility

After "Band Detection" all the facilities of Phoretix 1D are available in their standard form including: molecular size determination, Rf adjustments, band pattern matching, relationship dendrograms, wide range of tables and full reporting facilities.

7.6 Future Developments

Our efforts to date have focused mainly on categories of (PCR-based) genetic markers which can be set up for determination on short electrophoresis tracks. We have invested little or no effort yet in maximising throughput, in quantitative applications, RNA, protein, enzyme, or other applications. Genetic polymorphisms and approaches developed by us include:

1. Simple PCR checking gels and sizing gels (larger insertion/deletion polymorphisms) (>5% mobility difference resolution).
2. SNP analysis where RFLP is present or PCR-inducible (>5% mobility differences).
3. SNP analysis by ARMS-MADGE (192-well gels can be advantageous; see below).
4. Small insertion/deletion analysis by heteroduplex generator (HDG)-MADGE (192-well gels advantageous) (Voropanov and Day 2000).
5. De novo rare single base mutation (or SNP) scanning by Melt-MADGE (see above; Spanakis et al., submitted).
6. Resolution of <1.5% mobility differences using thermostatted MADGE with more attention to details of gel matrix and sample preparation. Applied so far to the resolution of 15 alleles of a minisatellite locus and to three tetranucleotide microsatellite polymorphisms, respectively, with 5, 7 and 17 alleles (O'Dell et al. 2000).
7. For higher throughput, 384-well MADGE using passive 96-channel loading and a dry gel system (Hinks et al., submitted).

Protocols for gel reuse (× 10) by DNase treatment and rebuffering (D. Palamand and I. Day, unpubl.) and many other variations (Table 7.2) have been developed but will not be considered further here. Original papers for points 3–7 listed above are in various stages of the publication process. In brief, at least three types of modified MADGE are useful.

7.6.1 High Resolution 96-Well MADGE

Attention to details (increased track length on 2 × 6 diagonal instead of 2 × 4; introduction of internal molecular weight markers in every track; sample dilu-

Table 7.2. Variations of MADGE used in our laboratory

Feature	Modification	Application
Matrix	Hydrolink	Higher resolution PA. More even 'pore size', also amphipathic. Reported application to heteroduplex resolution
	Page-Plus	Higher resolution PA. More even 'pore size'
	Duracryl	Mechanically robust PA which enables use of deep-pan PA MADGE without a glass backing, thus suitable for electroblotting (blot or purge for gel reuse), faster staining or if the glass represents an interference with imaging (see supports).
	Agaroses	For MW > 500 bp. Use with GelBond and/or as deep-pan. Agarose is much more mechanically robust than PA
Support	Lower fluorescence glass	Longer exposure/integration in conjunction with SYBR-Green
	Hydrophilic plastic (GelBond)	In conjunction with agarose, or PA. More ready detachment from gel (PA from Sticky Silane glass needs razor)
Format	H-PAGE	For horizontal PAGE with standard rows of wells rather than diagonal array. Simple setup otherwise as for MADGE
	Deeper/ thinner wells	$2 \times 2 \times 2$ mm is standard. Easy tip access, but open. Narrower wells permit less turbulence, but are less accessible to tips
	Deep-pan MADGE	Spacer gasket, e.g. 0.5 mm, used so that wells have gel at base. Prevents occasional interaction of sample with glass surface, and enables gel to be used unattached to solid support
	Stretch MADGE	Use of diagonals longer than the standard 2×4 (71.6°) angle. 2×6 under development for higher resolution analyses
Adapted PCRs	Shorter PCRs	PA MADGE makes convenient use and analysis of generally shorter PCRs (and often preferable for faster run times, etc.)
	CpG-PCR	Generates specific final common pathway of analysis (TaqI digest) for any CpG site
	Ultrashort PCRs	As for CpG-PCR, there is no difficulty with PA MADGE in analysing PCR products containing as few as one primer independent base pair. PA MADGE therefore leads to a different conceptual approach to some PCRs
Markers	Control tracks (rows outside the 96 array)	Useful for add in of post-PCR controls, but not compatible with 96-channel pipetting
	External reference points	A step toward automatic handling and analysis
	Internal markers in every track	Best approach for high-resolution applications

Table 7.2. *Continued.*

Feature	Modification	Application
Detection method	SYBR-Green and other fluors (intercalating or on a PCR oligo)	Greater detection sensitivity, or single-strand detection. Glass background fluorescence may require lower fluorescence glass, or (better) unbacked MADGE. If on the PCR oligo, then ensures that unincorporated label will not interfere with imaging of adjacent or more anodal tracks
	Silver staining	Shown to be possible, but little used (I. Day and R. Whittall, unpubl.).
	Radiolabels	Autoradiographic resolution generally problematic due to 2-mm-thick gels and short track lengths
Electrophoresis for de	Real-time-variable-temperature (Melt-MADGE)	Profiling of oligo dissociation by gel electrophoresis (complete ASO melting profiles). Duplex melting novo mutation scanning
	Transverse denaturing gradient DGGE H-PAGE and parallel gradient DGGE-MADGE	Under development for amplicon evaluation for Melt-MADGE (E.Spanakis and I. Day, unpubl.)
Using gels more than once	DNAase-treated gels	Gel treated serially with DNase, water/TE/EDTA washes, rebuffering. Compatible with glass-backed gels. Convenient for high-volume laboratory (D. Palamand and I. Day, unpubl.)
	Gel turned through 90° in electrophoresis tank	Requires sufficient track length to side as well as end of gel and wider gel tank. Bands are at right angles to original bands. Would be problematic when using software recognition. Most convenient reuse option for small-scale user
	Electroblotting perpendicular to array	Only possible with unbacked gels

SYBR-Green is from Molecular Probes, Oregon, USA, other useful fluorescence agents also from this source or (Vistra series) from Amersham International, Bucks, UK. Adaptations referenced * and general MADGE equipment and supplies are from MadgeBio Ltd., Grantham, UK. Hydrolink is from AT Baker, Duracryl from ESA Analytical Ltd., Cambs, UK, PagePlus from Amresco, Ohio, USA, and GelBond from FMC, Rockland, ME, USA

tion to avoid salt artefacts in the electrophoresis; use of better resolving acrylamide derivatives) enable resolution to less than 1.5% mobility differences of bands (Fig. 7.5). This has enabled calling of minisatellite and tetranucleotide repeat multiallelic microsatellite polymorphisms. We have been interested in

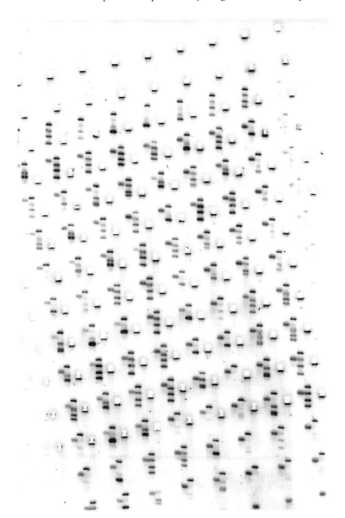

Fig. 7.5. High-resolution MADGE analysis of a multi-allelic microsatellite polymorphism. Products were run double stranded, so both homoduplexes and heteroduplexes are evident. In the heterozygote, the typical pattern is two resolved homoduplexes of different sizes, and a third lower mobility band in which the pair of heteroduplexes co-migrate. Mobilities are referenced to background mobility markers introduced in the sample loading buffer to every track. A five-allele system (15 genotypes) (*HUMTH01*) is shown

using such sites *within* genes as linkage disequilibrium markers in association studies and wider utility can be anticipated. In our system microsatellite amplicons are electrophoresed double stranded and there is the possibility that not only length polymorphic information but also internal sequence variation will be accessible by examining heteroduplex mobility.

7.6.2 192-Well MADGE

For checking one or two bands, much of the usual MADGE track length remains unused. One general approach to SNP calling is to use allele-specific PCR such as "ARMS" (amplification refractory mutation system, Newton et al. 1989): one simple robust format uses two separate allele-specific amplifications, one testing for each allele, with a control (unlinked) amplicon also included in each reaction as a positive control. In this case, it is easiest to load one gel and recombine the two tracks of information about a given template into one "virtual" software track. This can be readily achieved by locating a second well half way along the "original" MADGE track, giving two concatenated tracks of 12.25 mm (for 2 × 4 diagonal MADGE). This principle, convenient for software analysis and viewing, has also been applied to analysis of small insertion/deletion polymorphisms by heteroduplex analysis (A. Voropanov and I. Day, unpubl.). Post-PCR, heterozygotes are apparent from the reannealed PCR itself; homozygous types can be called by co-annealing with "generator" amplified from one or other homozygous types. The secondary (generator) assay is loaded in the second well of each virtual track (Fig. 7.6).

7.6.3 384-Well MADGE

384-Well microplates have become an established new standard in higher-throughput genomics laboratories for clone operations (arrayed libraries, gridding, spotting, storage, etc.) and for PCR. The difficulties in thin-walled plastics manufacture for the latter have been solved; 384-well microplates have wells in a 16 × 24, 4.5-mm pitch array. A diagonal turn of this array would result in MADGE wells and tracks too narrow (<1 mm) for manual access, but we have devised a 384-well MADGE format which combines four 96-well arrays in a linear (rather than tetradic) overlaid format (Fig. 7.7). The 1.5 × 1.5 × 1.5 mm wells achieved can be accessed by human hand. However, the array is too dense and confusing and undesirable for 8- or 12-channel loading, but 96 pin passive or air displacement transfer is not a problem for human hand and eye. The present approach to machining of the gel formers is near its limit because the space between "teeth" is small, but other arrangements of teeth, or other approaches to gel former production, may enable yet higher densities. However, it is likely that this would force the transition from economy of start up and versatility of the human operator, to expensive hardware configured around a core mini/microscale format and, at this level, complete departure from industry-standard microplate format may be appropriate.

7.6.4 The Next 5 Years

There is intense activity to systematise biological studies for many organisms, genome-wide clone analysis, genome-wide diversity studies, transcriptome

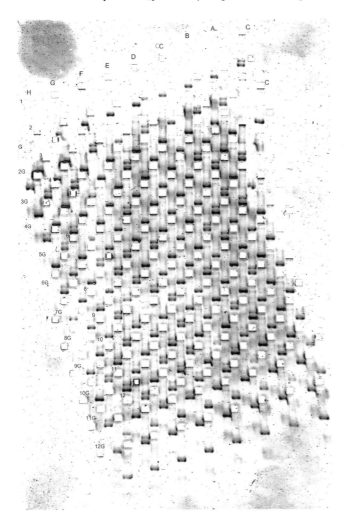

Fig. 7.6. Use of 192-well MADGE. In this application to heteroduplex analysis of a small (5 bp) insertion/deletion polymorphism, the second set of 96 wells is located half way along the tracks otherwise belonging to the usual set of 96 wells (see text). In this case, heterozygotes are recognised from heteroduplexes from the PCR amplicons loaded in the first set of wells; homozygotes for the rare allele are recognised from generation of heteroduplexes after co-annealing with 'generator' amplicons derived from a common allele homozygote, co-annealed products loaded in the second set of 96 wells. The two tracks for each template analysed are concatenated so that visual or software recombination of genotype information is simplified. The same approach is helpful also for two allele-specific amplifications for determination of single nucleotide polymorphism genotypes

and proteome studies and ensuant functional analyses. In every case, academic as well as commercial, there has been growing emphasis on throughput and complete description, in contrast to conjecture and hypothesis testing. A supplement to Nature Genetics "The Chipping Forecast", considered this situ-

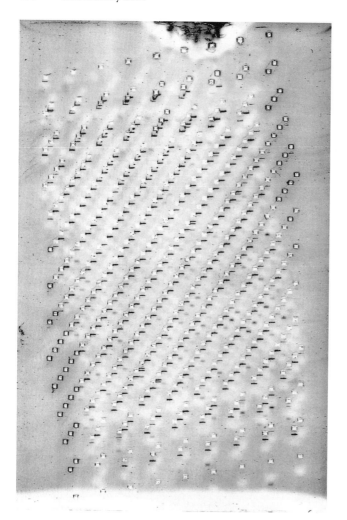

Fig. 7.7. Use of 384-well MADGE. Unlike the arrangement in 192-well microplates, the MADGE 384 wells are arranged in intercalated linear rather than intercalated tetradic format. This enables the use of larger ($1.5 \times 1.5 \times 1.5$ mm) wells, facilitating manual access as well as manufacture of gel formers and gels. In the example shown here, a rare allele usually observed in the heterozygous state can be recognised by the presence of two additional bands following a restriction enzyme digest. Visual track identification is possible with practice or with a template overlay, or, alternatively, software identification can be used. Electrophoresis loading time was 5 min, run time 15 min, followed by band detection. Using this system, 384-well PCR will remain the bottleneck

ation in detail (Lander 1999). There is no doubt that microchip arrays will become major information generators in these programs, using array densities from 10,000 to 1,000,000 grid positions. These systems, however, tend to lack end-user configurability, have restricted range of ready applicability such as nucleic acid hybridisations, require high cost hardware and have required substantial investments from the commercial sector. MADGE, drawing solely on the mature separation technology of electrophoresis, which is immediately applicable to many types of analysis, fills a junctional role in medium-through-put applications (100–10,000 data points per experiment). There is full end-user configurability and, often, immediate applicability with minimal development phases and minimal hardware needs. This reflects the origin of MADGE as the electrophoretic equivalent of the industry-standard microplate. We believe that MADGE applications will become very diverse, as have microplate applications. MADGE systems seem likely to occupy a range of niches in biomedical analyses which cannot be fulfilled by "chips" (assays not feasible or too expensive), where the throughput requirements are moderately high rather than very high, and where basic research requires immediate scale up. In our laboratory the advantages have been mainly for molecular genetic epidemiological studies of human disease, but it seems likely that any 96-well (or higher density) PCR block user could benefit from this system. Our research, however, has not explored the worlds of clone, protein, antibody and cell analyses. Application for construction of a high-density rat genome map (Watanabe et al. 1999) is a first example in clone analysis. We are aware (personal communications from various researchers) of groups trying to develop MADGE systems for protein, peptide, enzyme and other applications and others developing assay "services" based around MADGE. Further developments, ranging from basic research to establishment of product and applications to supply the research community, will be essential over the next 5 years for MADGE to fulfill its role as the intermediate between "old-fashioned" slab gels and "chips" upon which many of our future hopes in systematic research in the biological sciences rest.

Acknowledgements. INMD was previously a BHF Intermediate Fellow and is now a Lister Institute Research Fellow. The MRC is thanked for project grants G9605150 MB (ES), G9516890 MB (SO'D) and a ROPA award. SO'D is a Wessex Medical Trust Senior Research Fellow, A.V. is a Wessex Medical Trust PhD student. X.C. and L.H. are supported by the School of Medicine, University of Southampton. MADGE is subject to patent.

References

Bolla M, Haddad L, Winder AF, Humphries SE, Day INM (1995) High-throughput method for determination of apolipoprotein E genotypes with use of restriction digestion analysis by microplate array diagonal gel electrophoresis (MADGE). Clin Chem 41:1599–1604

Chee MS, Yang R, Hubbell E, Berno A, Huang XC, Stern D, Winkler J, Lockhart DJ, Morris MS, Fodor SPA (1996) Accessing genetic information in high-density DNA arrays. Science 274: 610–614

Clark SM, Mathies RA (1993) High-speed parallel separation of DNA restriction fragments using capillary array electrophoresis. Anal Biochem 215:163–170

Cooper DN, Krawczak M (1990) The mutational spectrum of single base-pair substitutions causing human genetic disease: patterns and predictions. Hum Genet 85:55–74

Cotton RGH (1997) Mutation detection. Oxford University Press, Oxford

Day INM, Humphries SE (1994a) Electrophoresis for genotyping: microtitre array diagonal gel electrophoresis (MADGE) on horizontal polyacrylamide (H-PAGE) gels, Hydrolink or agarose. Anal Biochem 222:389–395

Day INM, Humphries SE (1994b) Electrophoresis for genotyping: devices for high throughput using horizontal acrylamide gels (H-PAGE) and microtitre array diagonal gel electrophoresis (MADGE). Nature (product review) June 36–37

Day INM, Whittall R, Gundason V, Humphries SE (1995a) Dried template DNA, dried PCR oligonucleotides and mailing in 96-well plates: LDL receptor gene mutation screening. Biotechniques 18:981–984

Day INM, O'Dell SD, Cash ID, Humphries S, Weavind G (1995b) Electrophoresis for genotyping: temporal thermal gradient gel electrophoresis for profiling of oligonucleotide dissociation. Nucleic Acids Res 23:2404–2412

Day INM, Whittall R, O'Dell SD, Haddad L, Bolla M, Gudnason V, Humphries S (1997) Spectrum of LDL receptor gene mutations in heterozygous familial hypercholesterolaemia. Hum Mutat 10:116–127

Day INM, Palamand D, Spanakis E, Weavind GP, O'Dell SD (1998) Microplate array diagonal gel electrophoresis (MADGE) systems for molecular genetic epidemiology. Trends Biotechnol 16:287–290

Fischer SG, Lerman LS (1983) DNA fragments differing by single base-pair substitutions are separated in denaturing gradient gels: correspondence with melting theory. Proc Natl Acad Sci USA 80:1579–1583

Gelfi C, Cremonesi L, Ferrari M, Righetti PG (1996) Temperature-programmed capillary electrophoresis for detection of DNA point mutations. Biotechniques 21:926–928

Goffeau A (1997) DNA technology: molecular fish on chips. Nature 385:202–203

Guldberg P, Guttler F (1993) A simple method for the identification of point mutations using denaturing gradient gel electrophoresis. Nucleic Acids Res 21:61–62

Hacia JG, Brody LC, Chee MS, Fodor SPA, Collins FS (1996) Detection of heterozygous mutations in BRCA1 using high density oligonucleotide arrays and two-colour fluorescence analysis. Nat Genet 14:441–447

Hayashi K, Yandell DW (1993) How sensitive is SSCP? Hum Mutat 2:338–346

Henco K, Harders S, Wiese U, Riesner D (1994). Temperature gradient gel electrophoresis (TGGE) for the detection of polymorphic DNA and RNA. Methods Mol Biol 31:211–228

Kozulic B (1994). Looking at bands from another side. Anal Biochem 216:253–261

Lander ES (1999) Array of hope. Nat Genet 21 [Suppl 1]:3–4

Moyret C, Theillet C, Puig PL, Moles JP, Thomas G, Hamelin R (1994) Relative efficiency of denaturing gradient gel electrophoresis and single strand conformation polymorphism in the detection of mutations in exons 5 to 8 of the p53 gene. Oncogene 9:1739–1743

Newton CR, Graham A, Heptinstall LE, Powell SJ, Smith JC, Markham AC (1989) Analysis of any point mutation in DNA. The amplification refractory mutation system (ARMS). Nucleic Acids Res 17:2503–2516

O'Dell SD, Humphries SE, Day INM (1995) A rapid approach to genotyping of the insertion/deletion polymorphism in intron 16 of the angiotensin converting enzyme gene using simplified DNA preparation and microtitre array diagonal gel electrophoresis. Br Heart J 73:368–371

O'Dell SD, Humphries SE, Day INM (1996) PCR induction of a TaqI restriction site at any CpG dinucleotide using two mismatched primers (CpG-PCR). Genome Res 6:558–568

O'Dell SD, Miller GJ, Cooper JA, Hindmarsh PC, Pringle PJ, Ford H, Humphries SE, Day INM (1997) ApaI polymorphism in IGF2 gene and weight in middle-aged males. Int J Obesity 21:822–825

O'Dell SD, Chen X-h, Day INM (2000) High resolution microplate array diagonal gel electrophoresis: application to a multiallelic minisatellite. Hum Mutat 15:565–576

Parker LT, Zakeri H, Deng Q, Spurgeon S, Kwok PY, Niickerson DA (1996) AmpliTaq DNA polymerase, FS dye-terminator sequencing: analysis of peak height patterns. Biotechniques 21:
694–699

Risch N, Merikangas K (1996) The future of genetic studies of complex human diseases. Science
273:1516–1517

Scholz RB, Milde-Langosch K, Jung R, Schlechte H, Kabisch H, Wagener C, Loning T (1993) Rapid
screening for Tp53 mutations by temperature gradient gel electrophoresis: a comparison with
SSCP analysis. Hum Mol Genet 2:2155–2158

Sheffield VC, Beck JS, Kwitek AE, Sandstrom DW, Stone EM (1993) The sensitivity of single-strand
conformation polymorphism analysis for the detection of single base substitutions. Genomics
16:325–332

Studier FW (1973) Analysis of bacteriophage T7 early RNAs and proteins on slab gels. J Mol Biol
79:237–248

Sugden B, DeTroy B, Roberts RJ, Sambrook J (1975) Agarose slab-gel electrophoresis equipment.
Anal Biochem 68:36–46

Voropanov A, Day INM (2000) Elimination of dumbbell bands and poor resolution in MADGE
by use of delayed start electrophoresis. Biotechniques 28:32–4

Watanabe TK, Bihoreau M-T, McCarthy LC, Kiguwa SL, Hishigaki H, Tsuji A, Browne J, Yamasaki
Y, Miyakita AM, Oga K, Ono T, Okuno S, Kanemoto N, Takahashi E, Tomita K, Hayashi H,
Adachi M, Webber C, Davis M, Kiel S, Knights C, Smith A, Critcher R, Miller J, Thangarajah T,
Day PJR, Hudson JR, Irie Y, Takagi T, Nakamura Y, Goodfellow PN, Lathrop GM, Tanigami A,
James MR (1999) A radiation hybrid map of the rat genome containing 5255 markers. Nat
Genet 22:27–36

Wiese U, Wulpert M, Prusiner SB, Riesner D (1995) Scanning for mutations in the human prion
protein open reading frame by temporal temperature gradient gel electrophoresis. Electrophoresis 16:1851–1860

Whittall R, Gudnason V, Weavind G, Day LB, Humphries S, Day INM (1995) Utilities for high-
throughput use of the single strand conformational polymorphism method: screening of 791
patients with familial hypercholesterolaemia for mutations in exon 3 of the low density
lipoprotein receptor gene. J Med Genet 32:509–515

8 The Use of Sequence Analysis for Homozygote and Heterozygote Base Variation Discovery

Hans-Ulrich Thomann, Michael FitzGerald, Heidi Giese, and Kristen Wall

8.1 Introduction

The fluorescent version of Sanger dideoxy-sequencing has been in use for 15 years for gene discovery and comparative sequencing projects. Only a decade ago, detection of single-base changes, additions or deletions within genes of a given species were only possible provided that the organism was haploid or that the change occurred in the homozygous state. High background and the presence of significant peak height variation within bases of a given sequence read made it difficult to recognize when a base substitution occurred in the heterozygous state. In addition, the lack of efficient software to search for and call hetero- and homozygote single nucleotide polymorphisms (SNPs) provided a major obstacle to scaling up sequencing-based mutation/polymorphism discovery. However, with the advent of new sequencing chemistries, more accurate sequencing enzymes and automation-amenable SNP-scoring software, it is now possible to detect homozygous or heterozygous sequence variations with a high degree of sensitivity and accuracy. Thus, the combination of improved "wet lab" technology and sophisticated bioinformatics tools now allows for a more rigorous quantitative means of determining heterozygous and homozygous sequence polymorphisms.

In the following, we will elucidate the progress made in sequencing technologies and the impact on the identification of polymorphisms for disease gene discovery and validation as well as for pharmacogenomics. Automated fluorescent sequencing as well as various polymorphism-scoring software tools are described. We will also review different strategies utilized during polymorphism discovery, such as direct PCR sequencing (Genome Therapeutics Corporation) and subclone-mediated SNP discovery (SNP Consortium).

8.2 Sequence Analysis: a Polymorphism Discovery Method

One must distinguish *a priori* polymorphism detection, i.e., the discovery of unknown mutations or single nucleotide polymorphisms (SNPs) from

Principles and Practice
Molecular Genetic Epidemiology – A Laboratory Perspective
Ian N.M. Day (Ed.)
© Springer-Verlag Berlin Heidelberg 2002

polymorphism screening methods, i.e., the search or confirmation of known SNPs. The classic polymorphism discovery method is DNA sequence analysis (Nichols et al.1989; Tamary et al. 1994). However, during the last 10 years or so, enormous efforts have been made to search for alternatives to sequencing (reviewed in Landegren 1992; Cotton 1993, 1997; Taylor 1997) mostly due to the perceived high cost of this method and due to the lack of suitable automation.

Methods which will detect sequence polymorphisms in specific regions but will not reveal their nature or exact position are hereby termed class A SNP discovery methods. Typical representatives are: enzymatic mismatch scanning (EMS), single-stranded conformation polymorphism (SSCP), denaturing gradient gel electrophoresis (dGGE), and denaturing high performance liquid chromatography (dHPLC). Methods we hereby label as class B are capable of detecting and characterizing heterozygote and homozygote base changes at the same time. These criteria are to date only met by sequence analysis. Generally, once a polymorphism is detected via a class A method, sequence analysis must follow in order to confirm and characterize base variation(s).

Polymorphism screening methods are employed when searching for already known mutations or polymorphisms in well-defined loci. Examples are oligo-ligation assay (OLA) and ligase chain reaction (LCR), dideoxy fingerprinting (ddF) and mini-sequencing, allele-specific PCR (AS-PCR) and hybridization to low- or high-density gridded oligonucleotide targets. Several of these methods are reviewed in other chapters of this volume and are therefore not further analyzed in detail. Generally, discovery and screening methods are used in conjunction to maximize the efficiency of SNP detection.

Despite all efforts to find "stand alone" alternatives to sequence analysis, this technology is still considered the "gold standard" for heterozygote and homozygote polymorphism discovery. Only sequence analysis is proven to have a very low false negative rate during the discovery of new mutations and polymorphisms. This is especially critical when used in clinical studies. However, at this point, it must be mentioned that the accuracy and sensitivity of SNP discovery by sequence analysis are strongly dependent on the nature and quality of the template (see below).

8.3 Key Advancements in Sequencing Technologies: Chemistries, Hardware, and Software

8.3.1 Improvements in Sequence Analysis Technology

The discovery of thermostable DNA-dependent DNA polymerases (Saiki et al. 1986; Mullis et al. 1990a,b) was the first step toward improved and more efficient dideoxynucleotide-based sequence analysis (Sanger et al. 1977). It enabled the use of drastically reduced template amounts and formed the basis for higher efficiency through increased read lengths and improved sequence quality (Wong et al. 1987; Innis et al. 1988; Murray1989). The convergence of fluorescence-based detection technologies and PCR-based sequence analysis provided further support towards the rationalization and industrialization of sequence analysis (Kaiser et al. 1989; Wilson et al. 1990; Hunkapiller et al. 1991). DNA polymerases as well as detection chemistries have been continuously improved and are now sold as convenient kits or pre-mixes. Similar progress was achieved in separation and detection technologies. The turn of the millennium marked the final stage of the human genome project, a milestone in high-throughput sequence analysis, enabled mainly by the transition of slab-gel-based separation to capillary electrophoretic separation technology (reviewed by Dolnik1999). Molecular Dynamic's MegaBace 1000 (Amersham Pharmacia Biotech) represented the first off-the-shelf capillary device capable of separating the extension products of 96 sequence reactions in parallel. After its introduction in 1997, Perkin-Elmer (now Applied Biosystems) soon followed by introducing the ABI 3700 Prism. This device promised increased efficiency due to the ability to process four 384-well plates of reactions sequentially without manual intervention. In addition, after the introduction of linear polyacrylamide as separation matrix (Zhou et al. 2000), read lengths, which were thought to be only achievable via slab-gel devices, were also achievable on the MegaBace 1000.

8.3.1.1 Templates

Plasmid template-based high-throughput genome shotgun sequencing projects convinced us that template purity and quantity were important components to be considered in order to improve sequence quality and read lengths (Engelstein et al. 1998). These factors are also essential when newly introduced sequencing chemistries and separation technologies are tested and implemented into our process. Based on this experience, we have developed proprietary protocols to generate clean and defined templates for polymorphism discovery. The use of limiting amounts of template DNA such as genomic and cDNA or mitochondrial DNA for direct DNA sequence analysis after PCR-based amplification was described earlier (Wong et al. 1987; Wrischnik et al. 1987; Du et al. 1993; Chadwick et al. 1996). Others have used cloned templates, such as cosmid-inserts, to amplify and directly sequence specific regions for mutation analysis (Levedakou et al. 1989).

Our polymorphism discovery efforts begin with individual genomic DNAs isolated from blood or tissue samples. We then employ direct sequence analysis of PCR products. PCR fragments used as templates are generally 400–800 bp in length and are generated by nested or secondary PCR from very long, gene- or loci-specific amplification products (Fig. 8.1). This two-step amplification protocol ensures gene-specific polymorphism detection and avoids contamination of templates with co-amplified products from pseudogenes and

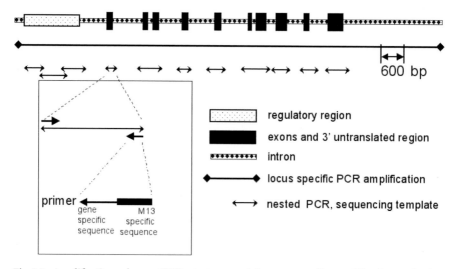

Fig. 8.1. Amplification scheme. GTC's strategy used for gene-specific amplification and subsequent template generation is shown. First, the drug target gene picked for this example is amplified via gene-specific primers, resulting in a 12.5-kb PCR product. This fragment serves as source DNA for nested PCR reactions to generate the sequencing templates covering important functional regions of the gene. The primers used for this PCR amplification are chimeric, fused at their 5′ end to M13 forward or M13 reverse primer sequences, respectively. The *inset* depicts the nested amplification and composition of a chimeric primer

highly homologous gene family members. In addition, this strategy proves to be very efficient as it allows the theoretical amplification (100% success rate) and sequence analysis of more than1000 loci from DNA isolated from only10 ml of total blood.

In order to facilitate efficient sequence analysis and to avoid primer-related problems, chimeric primers are used for a second, nested PCR amplification: 3′ portions of the forward and reverse primer are gene-specific; 5′ portions represent the M13 forward or reverse primer sequence, respectively (Fig. 8.1, inset). Thus, all templates synthesized by using these primers can be sequenced using standard M13 primers. This is especially important if dye primer chemistries are utilized, which are usually only available for use with priming sites found in common vectors. An additional advantage of using chimeric primers is the gain in sequencing efficiency as only two universal reaction mixes including primers need to be prepared and no primer-related reduction of the sequencing success rate needs to be taken into account.

As already mentioned, the length of the templates used for sequence analysis is chosen to be in the range of 400–800bp. This will ensure that by sequencing both strands of the fragment most of the coverage is double-stranded and of high-sequence quality (Fig. 8.2).

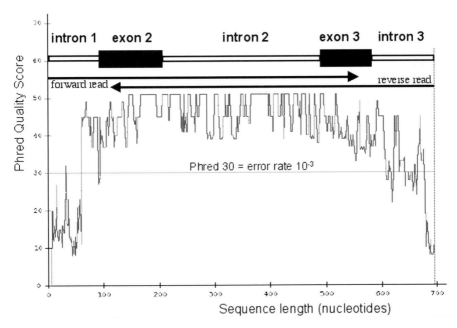

Fig. 8.2. Sequence quality and double-stranded sequence coverage. Following bidirectional sequence analysis, the quality of every base call is assessed using the software Phred (Ewing and Green 1998; Ewing et al.1998). Read mates, i.e., reads from opposite ends of one given template, are assembled in groups by the program Phrap (Washington University/Seattle) and a proprietary Assembly-Distributor algorithm developed at GTC. Typical mate reads from a 697-bp PCR fragment are shown, and the Phred quality assessment is graphically displayed. The PCR template covers two adjacent exons of 112- and 88-bp length, respectively, which are separated by an intron of 280 bp. Both reads each yield greater than 450 bases with a Phred score of higher than at least 30, i.e., an error rate of equal or less than 1 in 1000 bases (see line indicator). Bidirectional sequence analysis results in large, overlapping reads (>400 bp), providing high quality and in almost complete double-stranded coverage of functionally important regions

The accuracy of heterozygote base calling from sequences, which were obtained from diploid templates, is dependent on the faithful amplification of both alleles. Hare and Palumbi (1999) observed that in certain cases so-called null alleles, i.e., samples that are homozygote with respect to a wild-type or reference allele, can be a result of allele-biased PCR. Thus, it should be kept in mind that polymorphisms at the PCR priming site could influence bona fide representation of both alleles and obscure true heterozygotes. It has been suggested to confirm diploid-sequencing results using SNP screening methods on alternatively primed PCR products.

It should be mentioned that in many, but not all cases, allele-biased PCR can be detected if the distribution of homozygote and heterozygote genotypes does not follow the Hardy-Weinberg equilibrium (Hardy 1908).

8.3.1.2 Sequencing Chemistries

There are three main sequencing chemistries used for fluorescent, automated sequence analysis: dye primer chemistry, dye terminator chemistry, and internal dye label chemistry. Each of these chemistries has intrinsic properties, which render them more or less suited for polymorphism discovery, especially the discovery of heterozygote base variations. There are numerous examples in the literature validating each of these chemistries separately for use in polymorphism discovery. However, to date, no systematic comparison between chemistries and detection using commercially available platforms (ABI 377 Prism, MegaBace 1000, ABI 3700 Prism) has been published which elucidates the accuracy and sensitivity of each method and determines the "best combination" of chemistry and detection technology.

8.3.1.2.1 Dye Primer Label

Energy transfer (ET) dye primer chemistry (Amersham Pharmacia Biotech, Piscataway, NJ), which is used at Genome Therapeutics Corporation (GTC) for polymorphism discovery in positional cloning programs as well as for Drug Rescue™ programs, results in reads with exceptionally long stretches of high-quality base calls. ET primer chemistry is a second-generation dye primer chemistry, which takes advantage of an increased signal-to-noise ratio and therefore higher sensitivity during signal detection. Other dye primer sequencing kits, such as Big Dye primers, infrared dye primers and fluorescent dye primers, are commercially available from Applied Biosystems (Applera Corporation, Foster City, CA) and LiCOR (Lincoln, NE).

ET dye primer chemistry has been adapted for use in capillary electrophoresis instruments such as the MegaBace 1000 (Marsh et al. 1997; Kheterpal et al. 1996).

Although apparent difficulties associated with ET or dye primer chemistries can be easily overcome, they deserve to be noted:

1. Dye primer chemistry is sensitive to residual dNTPs left in the amplified template as well as to any nonspecific amplification products such as primer-dimer cogenerated during PCR amplification. Thus, template purification and/or modification of the PCR protocol in order to minimize the amount of detrimental contamination products left after amplification are important issues.
2. Dye primer sequencing demands separate reactions for each of the four bases, which can result in higher failure rates as compared to the "one tube" dye terminator reactions (see below) due to mistakes during reaction setup, evaporation during extension, and incomplete pooling of reactions. However, these problems can be minimized by automation. As we consider even peak heights and long read lengths an important factor for our SNP

discovery efforts, we have implemented ET dye primer sequencing into our SNP discovery process. We utilize 96-channel pipetting stations such as the Quadra (Tomtec, Hamden, CT) or the Hydra (Robbins Scientific, Sunnyvale, CA) to set up and pool reactions.

3. Strong sequencing stop sites, which can be due to secondary structure in the template and depurination of the template DNA, appear as "false termination" signals. Usually, these peaks are recognized as nonspecific termination because all four bases are represented in the signal peak. Thus, PCR conditions must be chosen carefully in order to avoid secondary structures in the template as well as to minimize depurination of the PCR template and products. False termination products usually carry a 3′ terminal deoxynucleotide, which allows the extension of these products with terminal deoxynucleotidyl transferase avoiding their detection during electrophoretic separation (Lasham and Darlison, 1993). However, this benefit is offset by the necessity to add additional steps to the sequencing protocol.

8.3.1.2.2 Dye Terminator Label

Compared with dye primer chemistries, dye terminator based sequencing protocols have seen dramatic improvements within the last 4 years. Uneven peak heights, a result of the DNA polymerase's differences in preference for each of the four fluorescently labeled dideoxynucleotides, were a distinct characteristic of terminator kits (Parker et al. 1995, 1996). Obviously, this trait had implications for polymorphism discovery: for instance, the low G signal observed when occurring after A might get repressed even further in the case of a heterozygote being present (e.g., G and C after A), thus giving the impression of a homozygote C at that position. However, new and improved polymerases as well as modifications to dye structures and buffer conditions have resulted in dramatic improvements in peak uniformity (Korch and Drabkin 1999a,b; Rosenblum et al. 1997; Zakeri et al. 1998). Big Dye Terminator chemistry (Applied Biosystems, Applera Corporation, Foster City, CA) is the most widely used sequencing chemistry for polymorphism discovery and generally accepted to be reliable and accurate. Dye terminator sequencing is usually more "forgiving" when regions with high G/C content or DNA secondary structures must be analyzed. Only correctly terminated extension products are detectable, thus this chemistry shows less background due to nonspecific termination. As mentioned above, nonspecific termination yields nonresolved and multiple dye peaks, which are encountered during dye primer sequencing.

In order to avoid nonspecific priming, which can cause high background, the amount of PCR primers left in the amplified template must be minimized. This can be achieved by adding a clean-up step or by adapting PCR conditions to reduce primer concentrations. In addition, residual dNTPs should be removed to avoid changing the ratios between dideoxy- and deoxynucleotides in the reaction kit, resulting in shorter read lengths or decreased signal-to-noise ratio during analysis.

During our polymorphism discovery efforts, we compared different sequencing chemistries, such as Amersham Pharmacia's ET Dye Primer and Applied Biosystem's Big Dye Terminator on older (ABI 377 Prism slab gel) and newer analysis platforms (MegaBace 1000 and ABI 3700 Prism capillary instruments).

8.3.1.2.3 Other Sequencing Chemistries and Labels

Numerous other fluorescence-based sequencing chemistries are commercially available. Some of them employ near-infrared labeling and are tied to specific electrophoresis devices such as the LI-COR (Lincoln, NE) IR^2 system (dye primer based detection on slab gels) or Beckman Coulter's (Fullerton, CA) CEQ 2000XL system (dye terminator based detection on eight-channel capillary device). Visible Genetics Inc. (Toronto, Canada) offers the OpenGene system, which was specifically designed for polymorphism discovery and screening through re-sequencing (Yager et al. 1999). The system requires use of Cy5 and Cy 5.5 labeled dye primer chemistry.

Yet another dye primer chemistry employs so-called Bodipy fluorescent dyes (Metzker et al. 1996). This chemistry is compatible with many commercially available detection devices such as the ABI 377 Prism and ABI 3700 Prism devices. The labels are very similar in chemical structure and do not require software-based mobility shift correction to interpret sequencing trace files.

Finally, a relatively young sequencing method, i.e., pyrosequencing, deserves mention (Ronaghi et al. 1998). Although this method is currently mainly used as a SNP screening method, it has potential as a class B SNP discovery method, provided that the read lengths can be improved from the present average of 50 bp. This method does not require any labels on either primer or deoxynucleotide. Instead, it detects incorporation of nucleotides into an extension product by a secondary assay via luciferase.

Under ideal conditions any sequencing protocol has the potential to produce high-quality data, if the workflow is fine-tuned and attributes the specific requirements of the sequencing chemistries and detection device used. When the emphasis lies on polymorphism discovery, one must also consider compatibility of peak height variations with the electrophoresis device's signal processing software as well as intrinsic characteristics of the base-calling algorithm used (see below). Careful comparison of new chemistries, separation technology, and base-calling software must proceed implementation to avoid problems with false negative or positive calls of heterozygote base exchanges.

8.3.1.3 Post-Sequence Treatment and Separation

Sequence reactions performed with Amersham Pharmacia's ET dye primer kits can be loaded directly after pooling of individual extension reactions onto the slab gels analyzed on the ABI 377 Prism. This is a convenient feature of energy-transfer dye-based detection, facilitated by the high signal-to-noise ratio of these dyes. Unfortunately, this "direct load" technique cannot be applied when capillary instruments are used for separation. This is an important trade-off to increased separation speed and ease of sample loading procedure, as these instruments do not tolerate salts and nonincorporated nucleotides.

In general, the efficiency of many clean-up methods lags behind the speed of sequencing and detection obtainable when modern capillary electrophoresis instruments are used. Clearly, when speed and efficiency count, one should not rely on methods involving single-spin columns or protocols involving multiple steps. Very often these methods are too dependent on individual skill and dedication. In addition, methods such as ethanol precipitation, which involve centrifugation steps, are not easy to automate. However, several commercially available post-sequencing clean-up kits in 96-well format are available, which are mostly based on size exclusion or ultrafiltration methods. In addition, modifications of template isolation methods can be used to prepare sequencing reactions prior to electrophoretic separation, such as magnetic- or glass-bead based protocols (Fangan et al. 1999 and Engelstein et al. 1998, respectively).

At GTC, we are developing a rapid desalting/nucleotide removal platform, the HT-MicroDialysis. The procedure is based on dialysis in microfiber capillaries. A clean-up cycle involves automatic injection of sequencing reactions into 96-well fiber cassettes, dialysis for 5–10 min and recovery in another 96-well plate for injection into capillary electrophoresis robots. This system is able to process submicroliter quantities and will soon be fully automated, eliminating any possibility of sample crossover during transfer from one reaction vessel to another.

8.4 Heterozygote and Homozygote Polymorphism Base Calling

A key technology within the sequence-based polymorphism discovery workflow is the SNP finding or SNP scoring algorithm. As mentioned at the beginning of the chapter, lack of software to search sequence traces for the presence of base alterations as compared to a reference sequence was a major obstacle for scaling-up SNP discovery by sequencing. Thus, class A discovery methods are used to pre-screen larger sample sizes for the presence of polymorphisms, which then in turn require confirmation and characterization by sequence analysis. However, as we stated at the beginning of this chapter, sequence-based SNP discovery *per se* is perfectly suited for high-throughput SNP discovery starting from a diploid DNA template. It will reveal not only the nature of the base variation, the location of the SNP and the state of zygosity, but also the frequencies of alleles all in one and the same step.

In the last 5 years great efforts have been made to develop a software package, which supports more or less automated recognition and calling of variations in a sequence as compared to a reference. Efficient algorithms should meet at least the first three of the following listed criteria:

1. Compatibility with the base-calling algorithm of the sequencing platform, i.e., support of trace file format yielded from the separation device.
2. Possibility of multiple sequencing read alignments against reference sequence. The software should allow the parallel display of reads covering the same sequence but originating from different individual DNAs.
3. Recognition and indication of base changes in individual reads as compared to a wild-type or reference sequence.
4. Quality-based sequence display, i.e., indication whether a possible base variation is located in a good quality sequence region or is more likely background and not to be called.
5. User-defined threshold ratio of reference or main peak versus secondary peak heights.

It is important to distinguish between algorithms that search for putative sequence variations based solely on ratios of main versus secondary peak, rather than algorithms that search for simultaneous main or reference peak height reduction and secondary peak height enlargement. It is clear that the latter will result in higher sensitivity as well as accuracy during searching and tagging of sequence variations. Below is a list of described software packages used during sequence variation discovery.

Factura (Applied Biosystems, Applera Corporation, Foster City, CA, USA). In order to support homo- and heterozygote base variation scoring, Applied Biosystems included "Factura" in the sequence comparison software package "Sequence Navigator"(Comparative PCR Sequencing: A Guide to Sequencing-based Mutation Detection. 1995, The Perkin-Elmer Corporation). Secondary peaks, as appearing in heterozygotes, can be detected and marked as an ambiguous base depending on the user-defined ratio threshold of any of the three "other" bases compared to the highest peak. Basically, criteria (1), (2), and (3) are met by the software, whereas (4) and (5) are not. Quality of sequence at and surrounding the putative base variation is not indicated and thus the certainty of the call is dependent on the user's judgement of the trace. In addition, the software will indicate putative heterozygotes even when the secondary peak is background, as it will not look for simultaneous reduction of the main peak.

Sequencher (Gene Codes Corporation, Ann Arbor, MI, USA). This sequence analysis software package combines its sequence alignment capability with base variation detection. Trace files, including those obtained from the major gel- as well as capillary-based separation platforms (ABI 377, ABI 3700, MegaBace 1000, etc.), can be viewed and aligned to those of other samples. A consensus sequence is displayed, indicating areas or single bases that differ in one or more of the individual reads. Heterozygote base variations are noted as ambiguities using the IUPAC naming convention. The program does not support quality-based assessment and thus cannot, without user intervention, distinguish real heterozygotes from false termination signals. The secondary peak triggers heterozygote marking but the primary peak variation is not taken into account.

PolyPhred (University of Washington, Seattle; Nickerson et al. 1997). This variation search and detection algorithm is a logical extension of the Phred/Phrap/Consed family of programs which were developed to support large-scale sequencing projects. Phred allows quality assessment of a base call based on peak symmetry, signal-to-noise, peak-peak separation, etc., and its prediction correlates well with the accuracy of final sequence assemblies (Ewing and Green 1998; Ewing et al. 1998; Richterich 1998). Phrap is used to assemble sequence reads based on regional overlaps. The user can choose force thresholds based on sequence quality, length of overlaps and penalties intro- duced by gaps as well as mismatches. Consed is a viewing tool for Phrap assem- blies that allows browsing of individual reads. PolyPhred compares individual sequence reads, and tags putative base variations. The user has the ability to choose thresholds based on sequence quality (default = Phred 30 or 1 error in 1000) and peak height ratios. The peak height ratio reflects a reduction of the main peak (default = reduction to at least 65% of original peak) and simulta- neous appearance of a secondary peak (default = increase to 25% or more rel- ative to background). Thus, PolyPhred fulfills all of the above-listed criteria. Newer versions also support export of information into databases. This includes a list of tagged bases, their coordinates (positions), and a confidence value (rank tags 1–6). The confidence value for consensus sequence tags increases if the SNP is tagged in read mates (forward and reverse read). It is also linked to the frequency of a specific base variation in a panel of reads (con- fidence increases with SNP frequency) and the presence of all three possible alleles in a given panel.

Trace Diff Algorithm (Bonfield et al. 1998). This software uses a trace differ- ence subtractor algorithm for variation detection. It compares experimental traces to a reference trace, which represents a consensus of electropherograms compiled from reads identified as reference. The software incorporates the Gap4 viewing tool. The software package fulfills criteria (1) to (5), as sequence quality is indicated using background gray scales. However, it should be mentioned that available documentation demonstrates only the detection of homozygote alterations, although heterozygote detection is mentioned briefly (http://www.mrc-lmb.cam.ac.uk/pubseq/).

During a large-scale study around the human ApoE locus, Lai et al. (1998) compared accuracy and sensitivity of three algorithms, i.e., the above-described Sequencher and PolyPhred and Lasergene (DNASTAR, Madison, WI), which was not mentioned. Within that group they identified PolyPhred as the most effective and least labor-intensive software, as it identified more polymorphisms than either of the other software packages. In similar large-scale projects, i.e., analysis of the human lipoprotein lipase, ACE and ApoE gene loci, Nickerson and coworkers have validated PolyPhred (Nickerson et al. 1998; Rieder et al. 1999; Nickerson et al. 2000, respectively). It is also suited to compare sequences that do not originate from diploid genomes such as mito-chondria (Rieder et al. 1998). At GTC, we have successfully used PolyPhred to screen plasmid clones for mutations, which were introduced by random mutagenesis methods.

A strong correlation between sequence quality and the ability to identify base variations, especially when occurring as heterozygotes, is obvious. If the quality is high, unambiguous identification of heterozygote and homozygote variations is achievable (Fig. 8.3). Putative hetero- and homozygote alleles are clearly identified and tagged by the software. However, depending on the thresholds defining the peak ratio (reduction of main peak/increase in secondary peak), "missed calls" are possible. We define missed calls as nontagged variations, which are manually detected when browsing through sequence alignments. The number of missed calls can vary depending on sequencing chemistry, initial base-calling algorithm of the separation platform, and sequence quality. Thus, we suggest careful examination of the required peak ratio values depending on the chemistry and platform used for SNP discovery. The setting should be chosen to support automatic tagging of all "real" variations, without inducing a dramatic increase in the number of tagged "false" alterations.

Bidirectional analysis of the region in question provides the best method to eliminate false-positive base variation calls. However, this is only valid if the sequence quality is satisfactory. The number of falsely tagged putative SNPs by PolyPhred decreased dramatically with increasing sequence quality as indicated by Phred (Nickerson et al. 1997).

8.5 Strategies for Sequencing-Based Polymorphism Detection

8.5.1 General Considerations

In theory it should be possible to detect sequence polymorphisms or muta-tions using any sequencing chemistry and separation/detection device. In practice there are important differences that must be taken into account when deciding which method is most suitable. First, high-quality sequence is essential if the goal is to discover novel polymorphisms as insufficient

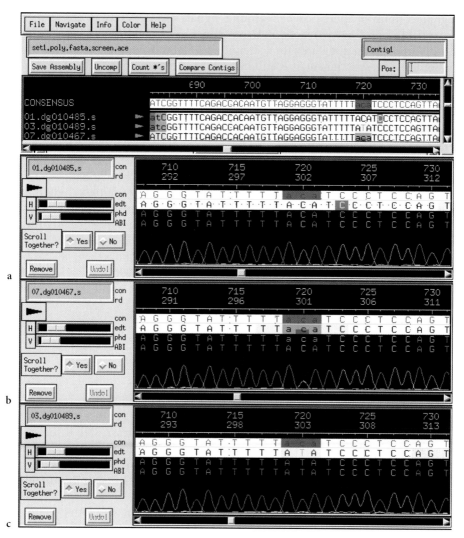

Fig. 8.3. Hetero- and homozygote search using PolyPhred. Sequencing traces aligned using PolyPhred (University of Washington, Seattle; Nickerson et al. 1997), represent all three possibilities of base variations: (a) homozygote allele wild-type (= reference sequence), (b) homozygote variant allele, and (c) heterozygote wild-type/variant allele. Sequence data were created by direct sequence analysis of PCR fragments using ET Dye Primer sequencing chemistry (Amersham Pharmacia Biotech). Sequence reaction products were analyzed on an ABI 377 Prism (Applied Biosystems, Applera Corporation) and after first pass base calling via the ABI 377 Prism specific software, chromatograms were called with Phred, assembled with Phrap, and examined for base variations by PolyPhred. The results are displayed using Consed. All software was licensed from the University of Washington, Seattle. The *upper window* shows the sequence alignment of three forward reads with a putative SNP indicated by the purple tag at position 720. A rough sequence quality is indicated by the shading of each (*white-gray-yellow* high-medium-low quality). Upon clicking on the highlighted position the *lower window* appears displaying the respective sequence traces

signal-to-noise or increased background signals can certainly interfere with correct calling of heterozygote and sometimes even homozygote base variations. Although the standards for screening loci for known polymorphisms can be somewhat less stringent, it is still important to minimize the rate of missed calls, and/or the number of false positives, which must be examined.

Sequence quality is dependent on many factors most notably the purity and integrity of the starting template, the choice of sequencing chemistry, the separation and base-calling method and, finally, the accuracy with which the protocol can be followed. Large sequencing operations, such as GTC's Sequencing Center, which processes thousands of samples in parallel, also benefit from automation as user-dependent fluctuations in quality can be avoided.

Nickerson and coworkers (Nickerson et al. 1997; Rieder et al. 1998) have pioneered the automation of polymorphism identification and scoring by creating the software PolyPhred. They have used this polymorphism-scoring algorithm in several projects together with direct bidirectional sequence analysis of PCR fragments covering large genomic regions (Boysen et al. 1996; Nickerson et al. 1998; Rieder et al. 1999; Nickerson et al. 2000).

We have adapted that strategy and modified, fine-tuned and integrated it into our high-throughput SNP Discovery platform (Fig. 8.1).

8.5.1.1 Starting Material and Study Goal

In order to support the goal of a polymorphism discovery study, sequencing templates must be chosen accordingly. If the study in question requires detection of polymorphisms in the coding regions of expressed genes, cDNA either directly or supplied as clone libraries can be used. In contrast, SNP discovery covering entire genes or loci must involve genomic DNA as starting material. If the goal is to discover polymorphisms in genomes of special populations or distinct species, genomic DNA can be processed, i.e., cloned into suitable vectors and supplied as libraries (see Sect. 8.5.3).

As already mentioned, the nature of nucleic acid chosen as starting material as well as the method selected for subsequent processing of the genetic material, such as cloning and PCR amplification, will greatly impact the SNP discovery process.

8.5.1.2 Sample Preparation and Sample Handling

Any biological material from which nucleic acids may be extracted is appropriate as a starting material for sequencing. There are two primary options for starting templates: (1) mRNA, which must first be converted via reverse transcription to cDNA, or (2) genomic DNA isolated from blood or tissue.

If mRNA, i.e., cDNA, is the starting material blood or tissue samples must be handled with much greater care as compared with samples from which DNA is isolated. Samples designated for RNA isolation must be collected in special vials avoiding contamination with RNases. Flash-freezing and storage at –70 °C are also mandatory, in addition to isolation protocols which preserve mRNA integrity.

DNA is much more stable. Although care should be taken during collection and storage, the primary concern is cross-contamination between individual samples rather than degradation. The use of cDNA as a starting template provides information on expressed genes only. There may also be a bias in the transcription of a wild-type gene such that heterozygosity is masked. In addition, recent experimental evidence supports the notion that the sensitivity of SNP detection may be significantly lower when cDNA is used as starting material compared to genomic DNA (Whatley et al. 1999).

In contrast to cDNA, by using genomic DNA the entire gene including noncoding key regions can be assessed. Because both alleles are equally represented in the sample, the likelihood of detecting any variations that are present in the heterozygous state is greatly improved.

Depending on the sequencing chemistry that is used, a PCR sample purification step may be required. This step is necessary to avoid problems due to primer or deoxynucleotide carryover into the sequencing reaction (see Sect. 8.3.1.3). Options might include dilution, ethanol precipitation, gel purification, and filtration (spin columns, etc.). A method frequently employed is the Exo/SAP protocol (Rieder et al. 1998). Unincorporated PCR primers and deoxynucleotide triphosphates are specifically degraded by exonuclease and shrimp alkaline phosphatase, respectively, allowing direct sequencing analysis without need for prior precipitation or other physical separation.

GTC's HT MicroDialysis technology (see Sect. 8.3.1.3) was designed to meet the needs of all clean-up steps required for direct PCR fragment-sequencing. Depending on the dialysis fiber size exclusion limit, it either allows removal of components interfering with capillary-based separation (salt, nucleotides) or components detrimental to sequencing reactions (carryover primers or deoxynucleotides).

8.5.2 SNP Discovery and Allele Frequency Detection Using Individual Genomic DNA Samples

8.5.2.1 General Considerations

If the goal of a SNP discovery study is to provide hetero- and homozygote base exchanges within a panel of individuals as well as their respective allele frequencies, the preferred starting material is genomic DNA isolated from individual blood or tissue samples.

Any method which will allow isolation of high molecular weight DNA from cells (e.g., Sambrook et al. 1989) can be used, including protocols supplied with commercially available kits, such as Qiagen's QIAamp blood kit (http://www.qiagen.com/) or Gentra Systems Inc.'s PUREGENE DNA isolation kit (http://www.gentra.com/) among many others. If large numbers of samples must be processed repeatedly, one might consider employing DNA isolation robots supplied by Gentra Systems Inc. (Autopure LS), Qiagen (http://www.qiagen.com/), Applera (http://www.appliedbiosystems.com/) and Autogen Inc. (AutoGen 740; http://www.autogen.com).

Mackey et al. (1998) described a method to isolate DNA from small quantities of blood and subsequent storage of the nucleic acid on cellulose paper. This method is gaining popularity when large libraries of individual DNAs must be stored. However, the DNA yield of this method might not be sufficient for sequencing-based SNP discovery of large genomic regions. This restriction in available material might also be relevant when the DNA must be isolated from archived tissue samples embedded in paraffin (Blomeke 1997). However, DNA isolated from these sources is generally not suited for long-range PCR amplification of specific loci (Fig. 8.1).

8.5.2.2 Sequencing Chemistry and Separation Platform

Table 8.1 provides a list of sequencing chemistries and their compatibility with commercially available separation devices. With so many options and even more possible combinations of chemistry and separation technology, it appears difficult to make the right choice when establishing a sequencing-based SNP discovery platform. Thus, it is important to establish a list of criteria and prioritize based on specific goals for each laboratory or organization. For example, one would probably not change platforms already established within a large-scale sequencing operation but seek to modify chemistries and protocols to ensure accuracy and sensitivity during SNP discovery. When building a new facility, however, a higher degree of freedom in choosing chemistry and platform exists. In this case, capillary electrophoresis technology is certainly the best option as it promises faster processing time, a higher degree of sample processing fidelity (no cross-contamination due to well loading as required with slab-gel technology) and the option for future technology upgrades and improvements.

GTC has made the transition from ABI 377 Prism slab-gel technology to capillary instruments. While most of the high-throughput analysis capacity is currently based on the MegaBace 1000 (Amersham Pharmacia Biotech), we also have ABI 3700 Prism (Applied Biosystems, Applera Corporation) capacity. However, we have maintained a number of ABI 377 platforms for certain SNP discovery projects, as this separation device in combination with ET dye primer chemistry consistently yields high-quality sequence over long read spans (>450 Phred 30 bases). For SNP discovery projects, which require the highest accuracy, sensitivity and tolerate no false negatives (SNPs not tagged by the software and thus missed), this proven platform ensures success.

Table 8.1. Sequencing chemistries and detection devices amenable for polymorphism discovery. Shown are commonly used labeling chemistries available in kits, compatible detection devices and their manufacturers

	Description	Sequencing chemistry or kit	Compatibility with separation and detection devices	Manufacturer
Energy transfer dye-terminator-based chemistries	Fluorescent Energy transfer dye label on the dideoxynucleotides	Big Dye Terminator	ABI 377 Prism, ABI 3700 Prism, ABI 3100 Prism, ABI 310 Prism	Chemistry & Device: Applied Biosystems (Applera Corporation), Foster City, CA, USA
	Fluorescent Energy transfer-based detection increases sensitivity and reduces background	ET Dye Terminator	MegaBace 1000, ABI 377 Prism	Chemistry & MegaBace 1000: Amersham Pharmacia Biotech, Piscataway, NJ, USA
Energy transfer dye-primer-based chemistries	Fluorescent Energy transfer dye label on the oligonucleotide primer	Big Dye Primer	ABI 377 Prism, ABI 3700 Prism, ABI 3100 Prism, ABI 310 Prism	Chemistry & Device: Applied Biosystems (Applera Corporation), Foster City, CA, USA
	Fluorescent Energy transfer-based detection increases sensitivity and reduces background	ET Dye Primer	MegaBace 1000, ABI 377 Prism	Chemistry & Megabace 1000: Amersham Pharmacia Biotech, Piscataway, NJ, USA
Dye primer based labels	Fluorescent dye label on the oligonucleotide primer	Infrared dye primers	LI-COR 7000	Chemistry & Device: LI-COR, Lincoln, NE, USA
		Cy5 or Cy5.5 dye primers	OpenGene System	Chemistry & Device: Visible Genetics Inc, Toronto, Canada
		Bodipy dye primers	ABI 377 Prism, ABI 3700 Prism	Dyes: NEN Life Sciences, Boston, MA, USA
Dye-terminator based labels	Fluorescent dye labels on the dideoxynucleotide	Near-infrared dye terminators	CEQ 2000XL system	Chemistry & Device: Beckman Coulter, Fullerton, CA, USA

In order to support increased throughput and efficiency during SNP discovery, we have tested dye primer and dye terminator chemistries on capillary electrophoresis platforms. Figure 8.4 shows sequence traces obtained with ET dye primer chemistry and separated on either gel- (ABI 377 Prism) or capillary-based (MegaBace 1000) devices. Both platforms support clear recognition of the heterozygote allele, when analyzed via PolyPhred, as indicated by the tags automatically assigned to the heterozygote allele in forward as well as reverse direction.

A comprehensive comparison looking at homo- and heterozygote variation discovery in relation to the sequence context as well as to the sequencing chemistry and separation platform employed is an ongoing development project at GTC. As already discussed in Section 8.3.1.2.2 (Dye Terminator Label), similar studies of first-generation dye terminator chemistries on the ABI 377 Prism platform showed significant peak height variations (Parker et al. 1995, 1996). However, even in the case of second-generation Big Dye Terminator chemistry, a recent study suggests that heterozygote alleles can be missed within a certain

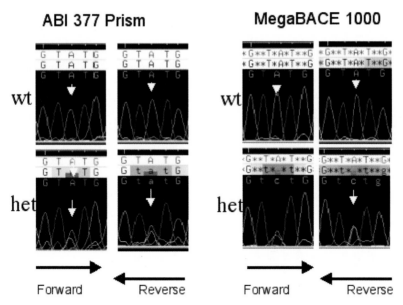

Fig. 8.4. Comparison of gel- and capillary-based platforms during SNP discovery. PCR fragments amplified from diploid DNA sources were sequenced bidirectionally using an ET Dye Primer kit (Amersham Pharmacia Biotech) and split to facilitate parallel separation on ABI 377 Prism and MegaBace 1000 devices. The sample designated for capillary electrophoretic separation was precipitated with ethanol and suspended in water prior to electrokinetic injection, while a third was used for direct load onto an ABI 377 gel as described by the kit manufacturer. After first pass base calling using the platform-specific base-calling algorithms, the traces were submitted to the Phred, Phrap, and PolyPhred pipeline and viewed with Consed (all software is licensed from the University of Washington, Seattle)

sequence context (Humma et al. 2000). We have encountered similar problems while separating Big Dye Terminator reactions on the ABI 377 Prism. Figure 8.5 illustrates one of these cases, where a heterozygote allele is recognized in both read directions and tagged accordingly by the software when sequenced with ET dye primer chemistry. In contrast, the same allele is not detectable in the reverse direction (neither by software nor while inspecting manually) when Big Dye Terminator chemistry was used for sequence analysis. In addition to those instances when heterozygote alleles are not recognized in one direction (Fig. 8.4), we also frequently observe absence of tags in both directions, mostly when sequence quality is not exceptionally high (>Phred 30 or less than 1 error in 1000). However, in these cases, the heterozygote allele is still recognized when manually scanning the traces for variations. SNPs that are discovered in one sequence read, but are apparently absent in the read mate, should be validated using an SNP screening method, especially if every discovered variation is important for the outcome of a study.

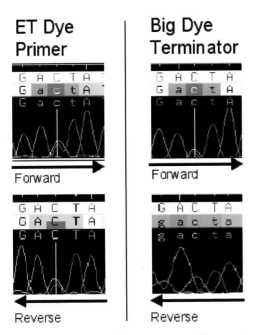

Fig. 8.5. Comparison of dye primer and dye terminator sequencing chemistry. PCR fragments amplified from diploid DNA sources were sequenced in parallel with either an ET Dye Primer kit (Amersham Pharmacia Biotech) or a Big Dye Terminator kit (Applied Biosystems, Applera Corporation). Reactions were precipitated with ethanol, suspended in loading dye according to the manufacturer's protocols, and analyzed on ABI 377 Prism devices. After first pass base calling using Applied Biosystem's base-calling algorithm, the traces were submitted to the Phred, Phrap, and PolyPhred pipeline and viewed with Consed

8.5.2.3 Automation and Data Tracking

Sequence-based SNP discovery requires a high degree of automation if many samples must be analyzed in parallel. For example, if the goal is to search for a SNP with an allele frequency of 5% and a confidence level of 95%, a minimum of 29 diploid templates originating from genomic DNA samples must be screened. Thus, if not supported by automation and data tracking tools, sequencing can be laborious and inefficient, especially if analysis of a locus or gene involves analysis of several fragments per individual (Fig. 8.1). Let us assume that SNP discovery on a locus requires amplification and analysis of ten different fragments to cover all functionally important regions, and that the combined PCR and sequencing success rate would be 90% for each of the ten fragments. Even at that exceptionally high success rate typically only 60% of all individuals are completely analyzed. Theoretically, complete coverage could be as low as 35% if PCR and/or sequencing failures were entirely random. Thus, in studies where SNP discovery over the entire locus is mandatory for each individual, failed samples must be identified, submitted for template re-amplification and/or re-sequencing and re-analyzed. At GTC we have adapted our "Finishing" platform for this process. Reads are continuously monitored and if a chosen quality threshold is reached and read mates (forward and reverse reads) are successful, the information is routed for PolyPhred analysis. Reads that do not reach the quality threshold, which is usually Phred 30 (1 error in 1000 base calls) over 65% of the template length (Fig. 8.2), are identified as failures.

Templates with two failing reads trigger re-amplification and subsequent sequence analysis request, whereas single read failure templates will be re-sequenced in the failed direction. A pipetting robot (Genesis 150 Workstation; TECAN, Durham, NC) is employed to assemble new work plates by reconfiguring primary PCR templates or sequencing templates, respectively.

8.5.3 Cloned DNA Templates (SNP Consortium)

As previously mentioned, several strategies can be used for the detection of SNPs via sequencing. A large, comprehensive SNP identification study is being conducted by the SNP Consortium, a nonprofit-making organization sponsored by ten large pharmaceutical companies, the Wellcome Trust (London, UK), IBM (Armonk, NY, USA), and Motorola Inc. (Schaumburg, IL, USA). The consortium's goal is to identify SNPs distributed throughout the human genome and make the data publicly available. Several large-scale sequencing centers, such as the Sanger Centre (Hinxton, UK), Washington University School of Medicine (St. Louis, MO, USA), the Whitehead Institute (Cambridge, MA, USA), the Cold Spring Harbor Laboratory (Cold Spring Harbor, NY, USA), and Stanford Human Genome Center (Palo Alto, CA, USA), are contributing to this effort. The project is expected to conclude at the end of 2001. In Fall 2000, more than 1.4 million unambiguous SNPs, over 800,000 mapped back to the human genome, were deposited in the database, thus overshooting the original goal and paving the path towards whole genome scanning using SNP markers (Altshuler et al. 2000; Mullikin et al. 2000; http://www.ncbi.nlm.nih.gov/SNP/).

The SNP discovery process starts with plasmid libraries generated using DNA pooled from several different individuals (9–24, depending on the center), all representing a unique ethnic group. Three different vectors are used, M13, pUC and pZERO2, to create libraries from human genomic DNA. In an early strategy termed "reduced representation subcloning" (RRS), restriction fragments obtained from genomic DNA pools were subcloned. More recently, and triggered by the available draft of the human genome, whole genome shotgun subcloning using randomly sheared fragments has replaced RRS. Clones are sequenced via dye terminator chemistry (Applied Biosystem's Big Dye Terminator kit) and analyzed on either slab-gel-based (ABI 377 Prism) or capillary-based (ABI 3700 Prism) automated sequencers. Resulting data are aligned and putative SNPs identified by comparing individual reads against the genome draft sequence using various automated programs including "BLAST" (http://snp.cshl.org/;).

All SNPs are subsequently validated by direct sequencing of PCR amplicons that have been purified by either solid-phase reversible immobilization or by shrimp alkaline phosphatase and exonuclease treatment (see Sect. 8.5.1.2.). During the validation phase either dye primers or dye terminators are used to sequence and samples are again analyzed using slab-gel-based or capillary-based automated sequencers.

As the goal of the SNP Consortium is to discover and catalogue human SNPs for use as markers for positional cloning of disease genes, disease marker detection and for whole genome scans, allele frequencies are not the primary focus. The frequency of alleles and hetero- and homozygotes in the genome can subsequently be determined by SNP scanning technologies, such as Affymetrix's GeneChip (www.affymetrix.com/), Sequenom's MassARRAY (www.sequenom.com/), Orchid's MegaSNPatron (www.orchid.com/) or GTC's patented HT-SNP assay, to mention only a few possibilities.

8.6 Summary and Outlook: To Sequence or Not To Sequence – This Is Not in Question

The general opinion that sequencing-based SNP discovery is nonefficient and expensive originates from the wrong assumption that sequence analysis has not evolved beyond gel, manual processing of samples and the missing support to find and record SNPs, like needles in a haystack, in a large number of trace files. Another reason to dismiss sequence analysis for SNP discovery frequently is inadequate economy of scale. In this chapter, we have tried to shed some light on the many advances in sequencing chemistries, automation and bioinformatics tools, which have made Sanger dideoxy-sequencing our method of choice for polymorphism discovery.

In Fig. 8.6, a schematic representation of GTC's SNP discovery workflow and its integration with SNP screening (validation of discovered SNPs) is shown. As soon as a gene or target region for SNP discovery is known, sequence databases (GenBank, HGTS, dbEST; all at http://www.ncbi.nlm.nih.gov/Database/index.html) are queried and a consensus is chosen from individual alignments. Functional regions are identified and mapped in order to support focused SNP discovery. In parallel, we use a proprietary protocol to identify SNPs in public and proprietary databases (e.g., dbSNP, dbEST, etc.), which map to the consensus gene sequence. Similar protocols and search algorithms for identifying SNPs in ESTs have been described previously (Garg et al.1999; Picoult-Newberg et al. 1999).

At this point gene-specific templates are generated by long-range PCR from individual genomic DNAs using gene-specific primer pairs. These PCR fragments are used to validate known SNPs using GTC's patented SNP screening assay (HT-SNP assay). They also serve as a source to generate sequencing templates (see Fig. 8.1).

The final product of a SNP discovery project is a detailed SNP map of the gene, including SNP frequencies with confidence intervals allowing a comparison between sample subgroups and the distribution of hetero- and homozygote alleles. The latter will subsequently allow the statistical prediction of haplotypes present in the samples examined, e.g., by using algorithms such as "Arlequin" (http://lgb.unige.ch/arlequin). If one takes into account that direct sequence analysis of diploid DNA templates will yield all the above information without requiring follow-up experiments, sequence-based SNP discovery can be considered economical and efficient. In other words, a fine-tuned sequencing-based SNP discovery protocol will save time and money as compared to SNP discovery involving SNP scanning methods (i.e., class A SNP discovery methods, see Sect. 8.2), which require subsequent sequencing of samples harboring putative polymorphisms.

Fig. 8.6. Polymorphism discovery and validation process diagram. Shown is the workflow used at GTC to discover sequence variations starting with individual genomic DNA and to validate proprietary and public SNPs. After choosing the gene or locus of interest, the "reference sequence" is assembled using information from public and proprietary databases. At the same time, published polymorphisms and information about regions important for gene function (coding sequences, splice sites and junctions, regulatory sequences, etc.) are mapped onto this sequence. Subsequently, primers are designed to amplify the entire gene locus (gene-specific amplification) and to generate sequencing templates covering all identified important regions. In the "Assay Development" mode, the gene-specific PCR as well as template generation are optimized. In the "Production Mode", the templates are primarily used for "SNP Discovery", i.e., they are submitted for bidirectional sequence analysis and subsequent SNP scoring using the software Phred, Phrap, PolyPhred, and Consed. However, they will also be available to validate newly discovered or published SNPs, which are recorded into the "SNP-DB"

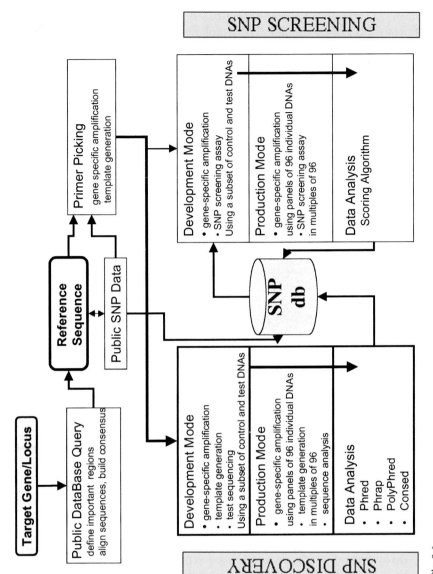

Fig. 8.6.

However, an efficient sequencing-based SNP discovery operation must seamlessly integrate automation of template synthesis, sequence analysis, and SNP finding/scoring in order to allow analysis of extended genes and loci in a larger number of samples (see Sect. 8.5.2.3). The operation should be scalable, allowing addition of robotics and sequencing devices to increase throughput without stretching the limits of data management and analysis. No less importantly it should allow changes in the workflow as well as instrumentation, in order to support timely integration of new or exchange of specific parts of the protocol.

Depending on the size of the operation, a laboratory information management system (LIMS) fully integrated with data analysis represents an important part of an efficient sequencing-based SNP discovery platform. A LIMS is strongly advisable if clinical samples must be processed, integrating bar-coded storage facilities, robotic workstations, reagents, and sample plates.

A possibility to increase efficiency and reduce analysis costs by 60% is presented by direct sequence analysis of pooled DNA samples but the degree of pooling is limited by the signal-to-noise ratio of the sequence trace (Taillon-Miller et al. 1999; Brouillette et al. 2000). When no false negatives are tolerated, e.g., in projects involving discovery of disease markers, pooling of templates will increase the risk of missing SNPs. However, when determining SNP frequencies in populations, this method is an alternative to a two-step discovery effort, involving scanning using a class A SNP discovery method followed by sequencing of selected samples.

New and improved sequencing chemistries are becoming available, which combine the advantages of dye terminator sequencing and more even peak heights usually observed during dye primer sequencing (e.g., Amersham Pharmacia's Thermosequenase II dye terminator kit), thus promising reduction of "missed calls" when employing automated SNP calling algorithms. A more general gain in sequencing efficiency can be expected when the next generation of capillary sequencers become commercially available (Scherer et al. 1999; Shi et al. 1999; Koutny et al. 2000). The increase in separation/detection speed must be accompanied by a more efficient front end processing, such as template generation and sequencing reactions, while drastically reducing sample volumes and thus cost. Single steps during a typical sequencing-based SNP discovery project, such as template synthesis and sequencing reactions, can be performed in capillary thermocycling devices reducing reaction volumes to less than 1 µl (Wittwer et al. 1989; Swerdlow et al. 1997). However, new developments in microfluidics, such as the "lab on chip" concept (Wells 1998), will soon be used to automate the entire workflow, beginning with blood or suitable tissue samples until separation of sequencing products using electrophoresis in micro devices (Tan and Yeung 1998; Bousse et al. 2000).

Keeping these improvements in mind, it is likely that sequencing-based polymorphism discovery will be the future method of choice.

Acknowledgments. We would like to thank Gary Breton, Shashi Prabhakar, and Michele Runge for assistance during data analysis and for many helpful discussions. Our sincere gratitude is extended to the dedicated team of the Sequencing Center and to Kim Fechtel and Hui Huang of the Bioinformatics Department at Genome Therapeutics Corporation.

References

Altshuler D, Pollara VJ, Cowles CR, Van Etten WT, Baldwin J, Linton L, Lander ES (2000) A SNP map of the human genome generated by reduced representation shotgun sequencing. Nature 407:513–516

Blomeke B, Bennett WP, Harris CC, Shields PG (1997) Serum, plasma and paraffin-embedded tissues as sources of DNA for studying cancer susceptibility genes. Carcinogenesis 18:1271–1275

Bonfield JK, Rada C, Staden R (1998) Automated detection of point mutations using fluorescent sequence trace subtraction. Nucleic Acids Res 26:3404–3409

Bousse L, Cohen C, Nikiforov T, Chow A, Kopf-Sill AR, Dubrow R, Parce JW (2000) Electrokinetically controlled microfluidic analysis systems. Annu Rev Biophys Biomol Struct 29:155–181

Boysen C, Carlson C, Hood E, Hood L, Nickerson DA (1996). Identifying DNA polymorphisms in human TCRA/D variable genes by direct sequencing of PCR products. Imunogenetics 44:121–127

Brouillette JA, Andrew JR, Venta PJ (2000) Estimate of nucleotide diversity in dogs with a pool-and-sequence method. Mamm Genome 11:1079–1086

Chadwick RB, Conrad MP, McGinnis MD, Johnston-Dow L, Spurgeon SL, Kronick MN (1996) Heterozygote and mutation detection by direct automated fluorescent DNA sequencing using a mutant Taq DNA polymerase. Biotechniques 20:676–683

Cotton RG (1993) Current methods of mutation detection. Mutat Res 285:125–144

Cotton RG (1997) Slowly but surely towards better scanning for mutations. Trends Genet 13:43–46

Dolnik V (1999) DNA sequencing by capillary electrophoresis. J Biochem Biophys Methods 41:103–119

Du Z, Hood L, Wilson RK (1993) Automated fluorescent DNA sequencing of polymerase chain reaction products. Methods Enzymol 218:104–121

Engelstein M, Aldredge TJ, Madan D, Smith JH, Mao JI, Smith DR, Rice PW (1998) An efficient, automatable template preparation for high throughput sequencing. Microb Comp Genomics 3:237–241

Ewing B, Green P (1998) Base-calling of automated sequencer traces using Phred. II. Error probabilities. Genome Res 8:186–194

Ewing B, Hillier L, Wendl MC, Green P (1998), Base-calling of automated sequencer traces using Phred. I. Accuracy assessment. Genome Res 8:175–185

Fangan BM, Dahlberg OJ, Deggerdal AH, Bosnes M, Larsen F (1999) Automated system for purification of dye-terminator sequencing products eliminates up-stream purification of templates. Biotechniques 26:980–983

Garg K, Green P, Nickerson DA (1999) Identification of candidate coding region single nucleotide polymorphisms in 165 human genes using assembled expressed sequence tags. Genome Res 9:1087–1092

Hardy GH (1908) Mendelian proportions in a mixed population. Science 28:49–50

Hare MP, Palumbi SR (1999) The accuracy of heterozygous base calling from diploid sequence and resolution of haplotypes using allele-specific sequencing. Mol Ecol 8:1750–1752

Humma LM, Farmerie WG, Wallace MR, Johnson JA (2000) Sequencing of beta 2-adrenoceptor gene PCR products using Taq BigDye terminator chemistry results in inaccurate base calling. Biotechniques 29:962–964, 966, 968

Hunkapiller T, Kaiser RJ, Koop BF, Hood L (1991) Large-scale DNA sequencing. Curr Opin Biotechnol 2:92–101

Innis MA, Myambo KB, Gelfand DH, Brow MA (1988) DNA sequencing with Thermus aquaticus DNA polymerase and direct sequencing of polymerase chain reaction-amplified DNA. Proc Natl Acad Sci USA 85:9436–9440

Kaiser RJ, MacKellar SL, Vinayak RS, Sanders JZ, Saavedra RA, Hood L (1989) Specific-primer-directed DNA sequencing using automated fluorescence detection. Nucleic Acids Res 17: 6087–6102

Kheterpal I, Scherer JR, Clark SM, Radhakrishnan A, Ju J, Ginther CL, Sensabaugh GF, Mathies RA (1996) DNA sequencing using a four-color confocal fluorescence capillary array scanner. Electrophoresis 17:1852–1859

Korch C, Drabkin H (1999a) Improved DNA sequencing accuracy and detection of heterozygous alleles using manganese citrate and different fluorescent dye terminators. Genome Res. 9:588–595

Korch C, Drabkin H (1999b) Manganese citrate improves base-calling accuracy in DNA sequencing reactions using rhodamine-based fluorescent dye-terminators. Nucleic Acids Res 27:1405–1407

Koutny L, Schmalzing D, Salas-Solano O, El-Difrawy S, Adourian A, Buonocore S, Abbey K, McEwan P, Matsudaira P, Ehrlich D (2000) Eight hundred-base sequencing in a microfabricated electrophoretic device. Anal Chem 72:3388–3391

Lai E, Riley J, Purvis I, Roses A (1998) A 4-Mb high-density single nucleotide polymorphism-based map around human APOE. Genomics 54:31–38

Landegren U (1992) Detection of mutations in human DNA. Genet Anal Tech Appl 9:3–8

Lasham A, Darlison MG (1993) Direct sequencing of lambda DNA from crude lysates using an improved linear amplification technique. Mol Cell Probes 7:67–73

Levedakou EN, Landegren U, Hood LE (1989) A strategy to study gene polymorphism by direct sequence analysis of cosmid clones and amplified genomic DNA. Biotechniques 7:438–442

Mackey K, Steinkamp A, Chomczynski P (1998) DNA extraction from small blood volumes and the processing of cellulose blood cards for use in polymerase chain reaction. Mol Biotechnol 9:1–5

Marsh M, Tu O, Dolnik V, Roach D, Solomon N, Bechtol K, Smietana P, Wang L, Li X, Cartwright P, Marks A, Barker D, Harris D, Bashkin J (1997) High-throughput DNA sequencing on a capillary array electrophoresis system. J Capillary Electrophor 4: 83–89

Metzker ML, Lu J, Gibbs RA (1996) Electrophoretically uniform fluorescent dyes for automated DNA sequencing. Science 271:1420–1422

Mullikin JC, Hunt SE, Cole CG, Mortimore BJ, Rice CM, Burton J, Matthews LH, Pavitt R, Plumb RW, Sims SK, Ainscough RM, Attwood J, Bailey JM, Barlow K, Bruskiewich RM, Butcher PN, Carter NP, Chen Y, Clee CM, Coggill PC, Davies J, Davies RM, Dawson E, Francis MD, Joy AA, Lamble RG, Langford CF, Macarthy J, Mall V, Moreland A, Overton-Larty EK, Ross MT, Smith LC, Steward CA, Sulston JE, Tinsley EJ, Turney KJ, Willey DL, Wilson GD, McMurray AA, Dunham I, Rogers J, Bentley DR (2000) An SNP map of human chromosome 22. Nature 407:516–520

Mullis (1990a) The unusual origin of the polymerase chain reaction. Sci Am 262:56–61, 64–65

Mullis (1990b) Target amplification for DNA analysis by the polymerase chain reaction. Ann Biol Clin 48:579–582

Murray V (1989) Improved double-stranded DNA sequencing using the linear polymerase chain reaction. Nucleic Acids Res 17:8889

Nichols WC, Liepnieks JJ, McKusick VA, Benson MD (1989) Direct sequencing of the gene for Maryland/German familial amyloidotic polyneuropathy type II and genotyping by allele-specific enzymatic amplification. Genomics 5:535–540

Nickerson DA, Tobe VO, Taylor SL (1997) PolyPhred: automating the detection and genotyping of single nucleotide substitutions using fluorescence-based resequencing. Nucleic Acids Res 25:2745–2751

Nickerson DA, Taylor SL, Weiss KM, Clark AG, Hutchinson RG, Stengard J, Salomaa V, Vartiainen E, Boerwinkle E, Sing CF (1998) DNA sequence diversity in a 9.7-kb region of the human lipoprotein lipase gene. Nat Genet 19:233–240

Nickerson DA, Taylor SL, Fullerton SM, Weiss KM, Clark AG, Stengard JH, Salomaa V, Boerwinkle E, Sing CF (2000) Sequence diversity and large-scale typing of SNPs in the human apolipoprotein E gene. Genome Res 10:1532–1545

Parker LT, Deng Q, Zakeri H, Carlson C, Nickerson DA, Kwok PY (1995) Peak height variations in automated sequencing of PCR products using Taq dye-terminator chemistry. Biotechniques 19:116–121

Parker LT, Zakeri H, Deng Q, Spurgeon S, Kwok PY, Nickerson DA (1996) AmpliTaq DNA polymerase, FS dye-terminator sequencing: analysis of peak height patterns. Biotechniques 21:694–699

Picoult-Newberg L, Ideker TE, Pohl MG, Taylor SL, Donaldson MA, Nickerson DA, Bayce-Jacino M (1999) Mining SNPs from EST databases. Genome Res 9:167–174

Richterich P (1998) Estimation of errors in "raw" DNA sequences: a validation study. Genome Res 8:251–259

Rieder MJ, Taylor SL, Tobe VO, Nickerson DA (1998) Automating the identification of DNA variations using quality-based fluorescence re-sequencing: analysis of the human mitochondrial genome. Nucleic Acids Res 26:967–973

Rieder MJ, Taylor SL, Clark AG, Nickerson DA (1999) Sequence variation in the human angiotensin converting enzyme. Nat Genet 22:59–62

Ronaghi M, Uhlen M, Nyren P (1998) A sequencing method based on real-time pyrophosphate. Science 281:363–365

Rosenblum BB, Lee LG, Spurgeon SL, Khan SH, Menchen SM, Heiner CR, Chen SM (1997) New dye-labeled terminators for improved DNA sequencing patterns. Nucleic Acids Res 25:4500–4504

Saiki RK, Bugawan TL, Horn GT, Mullis KB, Erlich HA (1986) Analysis of enzymatically amplified beta-globin and HLA-DQ alpha DNA with allele-specific oligonucleotide probes. Nature 324:163–166

Sambrook J, Fritsch EF, Maniatis T (1989) Molecular cloning: a laboratory manual. Cold Spring Harbor Lab Press, Cold Spring Harbor

Sanger F, Nicklen S, Coulson AR (1977) DNA sequencing with chain-terminating inhibitors. Proc Natl Acad Sci USA 74:5463–5467

Scherer JR, Kheterpal I, Radhakrishnan A, Ja WW, Mathies RA (1999) Ultra-high throughput rotary capillary array electrophoresis scanner for fluorescent DNA sequencing and analysis. Electrophoresis 20:1508–1517

Shi Y, Simpson PC, Scherer JR, Wexler D, Skibola C, Smith MT, Mathies RA (1999) Radial capillary array electrophoresis microplate and scanner for high-performance nucleic acid analysis. Anal Chem 71:5354–5361

Swerdlow H, Jones BJ, Wittwer CT (1997) Fully automated DNA reaction and analysis in a fluidic capillary instrument. Anal Chem 69:848–855

Taillon-Miller P, Piernot EE, Kwok PY (1999) Efficient approach to unique single-nucleotide polymorphism discovery. Genome Res 9:499–505

Tan H, Yeung ES (1998) Automation and integration of multiplexed on-line sample preparation with capillary electrophoresis for high-throughput DNA sequencing. Anal Chem 70:4044–4053

Tamary H, Surrey S, Kirschmann H, Shalmon L, Zaizov R, Schwartz E, Rappaport EF (1994) Systematic use of automated fluorescence-based sequence analysis of amplified genomic DNA for rapid detection of point mutations. Am J Hematol 46:127–133

Taylor GR (ed) (1997) Laboratory methods for the detection of mutations and polymorphisms in DNA. CRC Press, Boca Raton

Wells WA (1998) The next chip-based revolution. Chem Biol 5:R115-R116

Whatley SD, Woolf JR, Elder GH (1999) Comparison of complementary and genomic DNA sequencing for the detection of mutations in the HMBS gene in British patients with acute intermittent porphyria: identification of 25 novel mutations. Hum Genet 104:505–510

Wilson RK, Chen C, Avdalovic N, Burns J, Hood L (1990) Development of an automated procedure for fluorescent DNA sequencing. Genomics 6:624–634

Wittwer CT, Fillmore GC, Hillyard DR (1989) Automated polymerase chain reaction in capillary tubes with hot air. Nucleic Acids Res 11:4353–4257

Wong C, Dowling CE, Saiki RK, Higuchi RG, Erlich HA, Kazazian HH Jr (1987) Characterization of beta-thalassaemia mutations using direct genomic sequencing of amplified single copy DNA. Nature 330:384–386

Wrischnik LA, Higuchi RG, Stoneking M, Erlich HA, Arnheim N, Wilson AC (1987) Length mutations in human mitochondrial DNA: direct sequencing of enzymatically amplified DNA. Nucleic Acids Res 15:529–542

Yager TD, Dunn JM, Stevens JK (1999) High-speed DNA sequencing in ultrathin slab gels. Curr Opin Biotech 8:107–113

Zakeri H, Amparo G, Chen SM, Spurgeon S, Kwok PY (1998) Peak height pattern in dichlororhodamine and energy transfer dye terminator sequencing. Biotechniques 25:406–410, 412–414

Zhou H, Miller AW, Sosic Z, Buchholz B, Barron AE, Kotler L, Karger BL (2000) DNA sequencing up to 1300 bases in two hours by capillary electrophoresis with mixed replaceable linear polyacrylamide solutions. Anal Chem 72:1045–1052

Subject Index

Printing (Computer to Film): Saladruck Berlin
Binding: Stürtz AG, Würzburg

DATE DUE